WASTE MINIMIZATION DURING THE LIFE CYCLE OF NUCLEAR POWER PLANTS

The following States are Members of the International Atomic Energy Agency:

AFGHANISTAN
ALBANIA
ALGERIA
ANGOLA
ANTIGUA AND BARBUDA
ARGENTINA
ARMENIA
AUSTRALIA
AUSTRIA
AZERBAIJAN
BAHAMAS
BAHRAIN
BANGLADESH
BARBADOS
BELARUS
BELGIUM
BELIZE
BENIN
BOLIVIA, PLURINATIONAL STATE OF
BOSNIA AND HERZEGOVINA
BOTSWANA
BRAZIL
BRUNEI DARUSSALAM
BULGARIA
BURKINA FASO
BURUNDI
CABO VERDE
CAMBODIA
CAMEROON
CANADA
CENTRAL AFRICAN REPUBLIC
CHAD
CHILE
CHINA
COLOMBIA
COMOROS
CONGO
COOK ISLANDS
COSTA RICA
CÔTE D'IVOIRE
CROATIA
CUBA
CYPRUS
CZECH REPUBLIC
DEMOCRATIC REPUBLIC OF THE CONGO
DENMARK
DJIBOUTI
DOMINICA
DOMINICAN REPUBLIC
ECUADOR
EGYPT
EL SALVADOR
ERITREA
ESTONIA
ESWATINI
ETHIOPIA
FIJI
FINLAND
FRANCE
GABON
GAMBIA, THE
GEORGIA
GERMANY
GHANA
GREECE
GRENADA
GUATEMALA
GUINEA
GUYANA
HAITI
HOLY SEE
HONDURAS
HUNGARY
ICELAND
INDIA
INDONESIA
IRAN, ISLAMIC REPUBLIC OF
IRAQ
IRELAND
ISRAEL
ITALY
JAMAICA
JAPAN
JORDAN
KAZAKHSTAN
KENYA
KOREA, REPUBLIC OF
KUWAIT
KYRGYZSTAN
LAO PEOPLE'S DEMOCRATIC REPUBLIC
LATVIA
LEBANON
LESOTHO
LIBERIA
LIBYA
LIECHTENSTEIN
LITHUANIA
LUXEMBOURG
MADAGASCAR
MALAWI
MALAYSIA
MALI
MALTA
MARSHALL ISLANDS
MAURITANIA
MAURITIUS
MEXICO
MONACO
MONGOLIA
MONTENEGRO
MOROCCO
MOZAMBIQUE
MYANMAR
NAMIBIA
NEPAL
NETHERLANDS, KINGDOM OF THE
NEW ZEALAND
NICARAGUA
NIGER
NIGERIA
NORTH MACEDONIA
NORWAY
OMAN
PAKISTAN
PALAU
PANAMA
PAPUA NEW GUINEA
PARAGUAY
PERU
PHILIPPINES
POLAND
PORTUGAL
QATAR
REPUBLIC OF MOLDOVA
ROMANIA
RUSSIAN FEDERATION
RWANDA
SAINT KITTS AND NEVIS
SAINT LUCIA
SAINT VINCENT AND THE GRENADINES
SAMOA
SAN MARINO
SAUDI ARABIA
SENEGAL
SERBIA
SEYCHELLES
SIERRA LEONE
SINGAPORE
SLOVAKIA
SLOVENIA
SOMALIA
SOUTH AFRICA
SPAIN
SRI LANKA
SUDAN
SWEDEN
SWITZERLAND
SYRIAN ARAB REPUBLIC
TAJIKISTAN
THAILAND
TOGO
TONGA
TRINIDAD AND TOBAGO
TUNISIA
TÜRKİYE
TURKMENISTAN
UGANDA
UKRAINE
UNITED ARAB EMIRATES
UNITED KINGDOM OF GREAT BRITAIN AND NORTHERN IRELAND
UNITED REPUBLIC OF TANZANIA
UNITED STATES OF AMERICA
URUGUAY
UZBEKISTAN
VANUATU
VENEZUELA, BOLIVARIAN REPUBLIC OF
VIET NAM
YEMEN
ZAMBIA
ZIMBABWE

The Agency's Statute was approved on 23 October 1956 by the Conference on the Statute of the IAEA held at United Nations Headquarters, New York; it entered into force on 29 July 1957. The Headquarters of the Agency are situated in Vienna. Its principal objective is "to accelerate and enlarge the contribution of atomic energy to peace, health and prosperity throughout the world".

IAEA-TECDOC-2091

WASTE MINIMIZATION DURING THE LIFE CYCLE OF NUCLEAR POWER PLANTS

INTERNATIONAL ATOMIC ENERGY AGENCY
VIENNA, 2025

COPYRIGHT NOTICE

Publishing Section
International Atomic Energy Agency
Vienna International Centre
PO Box 100
1400 Vienna, Austria
tel.: +43 1 2600 22529 or 22530
email: sales.publications@iaea.org
www.iaea.org/publications

For further information on this publication, please contact:

Waste Technology Section
International Atomic Energy Agency
Vienna International Centre
PO Box 100
1400 Vienna, Austria
Email: Official.Mail@iaea.org

Printed by the IAEA in Austria
April 2025
https://doi.org/10.61092/iaea.vfpq-k3z1

IAEA Library Cataloguing in Publication Data

Names: International Atomic Energy Agency.
Title: Waste minimization during the life cycle of nuclear power plants / International Atomic Energy Agency.
Description: Vienna : International Atomic Energy Agency, 2025. | Series: IAEA TECDOC series, ISSN 1011-4289 ; no. 2091 | Includes bibliographical references.
Identifiers: IAEAL 25-01753 | ISBN 978-92-0-107725-7 (paperback : alk. paper) | ISBN 978-92-0-107825-4 (pdf)
Subjects: LCSH: Radioactive wastes. | Radioactive wastes — Management. | Radioactive wastes — Storage. | Radioactive waste disposal.

FOREWORD

Radioactive waste produced during the operation of nuclear power plants requires careful management to ensure both safety and cost effectiveness. In general, Member States with a nuclear programme have incorporated the requirement of radioactive waste minimization. Although safety requirements for the management of waste from nuclear power plants need to be fulfilled unconditionally, the implementation of cost efficient processing depends on the radioactive waste management and spent fuel management policies adopted in a particular Member State. The policies and frameworks for radioactive waste management and spent fuel management vary by country, influenced by regulatory environments, economic considerations and public acceptance.

The waste produced can be categorized into several types, such as low level waste, intermediate level waste, high level waste and spent fuel. Each category has specific handling, processing and disposal requirements to mitigate environmental and health risks. The cost efficient processing of radioactive waste is critical to the economic viability of nuclear power, and the strategies to reduce costs include waste volume minimization, recycling and reprocessing of materials, and the development of more efficient waste treatment technologies. This publication provides end users with technical information regarding waste minimization at nuclear power plants based on the experiences of nuclear power plant operators. It focuses on typical waste inventories, minimization methodologies, the optimization of waste minimization programmes and the evaluation of practices.

Effective waste management in nuclear power plants is a balancing act between ensuring safety and achieving cost efficiency. By adopting comprehensive waste minimization strategies and leveraging technical advancements, nuclear power plant operators can manage radioactive waste responsibly. The publication serves as a valuable resource, providing practical insights and strategies to help nuclear power plant operators optimize their waste management practices, ultimately contributing to the sustainable operation of nuclear power plants.

This publication was developed with the valuable assistance of the international experts listed at the end of the publication. The IAEA officers responsible for this publication were W. Meyer and S. Janski of the Division of Nuclear Fuel Cycle and Waste Technology.

EDITORIAL NOTE

CONTENTS

1. INTRODUCTION

1.1. BACKGROUND

During the operation of nuclear power plants (NPPs), are generated radioactive waste that requires careful management to ensure both safety and cost-effectiveness. The IAEA Safety and Security Glossary [1] defines waste minimization as: "The process of reducing the amount and activity of radioactive waste to a level as low as reasonably achievable, at all stages from the design of a facility or activity to decommissioning, by reducing waste generation and by means such as recycling and reuse, and treatment, with due consideration for secondary as well as primary waste". Article 11 of the Joint Convention on the Safety of Spent Fuel Management states that Contracting parties needs to take the appropriate steps to minimize the generation of radioactive waste [2].

Over the past few decades, NPP operators have continued to build on their experience in the minimization of waste arisings thereby accumulating experience in cost effectively minimizing (optimizing) of the waste [3]. This publication describes these approaches and processes to help other programmes in further improving their own waste minimization strategy during the life cycle of NPPs.

1.2. OBJECTIVE

The objective of this publication is to provide Member States with information and examples for the minimization of radioactive waste volumes during the life cycle of NPPs. The publication serves as a valuable resource, providing practical insights to help NPP operators optimize their waste management practices, ultimately contributing to the sustainable operation of nuclear power plants.

1.3. SCOPE

This publication contains information regarding methodologies, practices, and approaches to minimize radioactive waste volumes during the operation of NPPs. The main focus of the publication deals with the technical aspects but is widely recognized that the non-technical issues are also important (e.g., policies and strategies, work optimization, administrative rules and internal procedures) and therefore they are briefly addressed. The provided information could facilitate the planning and designing of facilities and decommissioning activities, it is believed that the publication will constitute a support for the NPP operators, planners, and implementers.

1.4. STRUCTURE

This publication provides comprehensive insights into the strategies for minimizing waste volumes generated from the operations of Nuclear Power Plants (NPPs). It serves as a guide for readers, outlining the various factors that impact waste minimization strategies for different waste streams, as well as the methodologies employed, and the processes used for monitoring these initiatives. The publication is structured as follows:

Section 2 delves into the critical factors influencing waste minimization management. This section explores regulatory frameworks, operational practices, technological advancements, and safety considerations that shape the approach to waste minimization. It emphasizes the importance of understanding these factors to develop effective strategies tailored to specific waste streams.

Chapters 3, 4, and 5 are dedicated to presenting minimization strategies for liquid waste, wet solid waste, and dry solid waste, respectively. Each chapter outlines the unique characteristics of these waste types, the challenges associated with their management, and the innovative practices being implemented to reduce their volumes. Detailed case studies illustrate successful approaches and highlight best practices in the field.

In Section 6, the publication addresses the influence of metal composition on waste generation. This section provides an in-depth discussion on how different materials used in NPP operations contribute to waste profiles. It examines the impact of material selection on waste minimization, and strategies for optimizing metal usage to reduce overall waste generation.

Section 7 presents an overview of the current methodologies used to monitor, assess, and benchmark waste minimization programmes. This section includes a discussion of various assessment tools, performance indicators, and best practices for evaluating the effectiveness of waste management strategies. It emphasizes the importance of continuous improvement and the need for robust monitoring frameworks to ensure that waste minimization efforts are meeting regulatory and environmental standards.

The main conclusions drawn at the end of the main body of the publication summarize the key findings and recommendations for enhancing waste minimization practices. The publication also includes annexes that feature practical examples from various Member States, showcasing successful initiatives, lessons learned, and innovative solutions to common challenges in waste management.

2. WASTE MIMIMIZATION CONSIDERATIONS

One of the objectives of a waste management programme is to minimize waste volume (m^3), weight (kg) and activity (Bq) of radioactive waste, to reduce the financial burden on waste management activities (e.g., processing, storage, and disposal). For example, during midlife replacement (MLR) of a reactor core, the activity of the near-core structures and components that could measure up to hundreds of thousands of TBq/kg. Waste minimization activities could reduce this activity, facilitate processing, and will directly reduce the impact of disposed waste on the environment.

2.1. OVERVIEW OF MINIMIZATION STRATEGIES

The radioactive waste generated during the NPP's lifetime operation can be controlled in terms of volume, weight and activity by implementing various measures, procedures and protocols, such as:

- Implementing dedicated waste minimization programmes;
- Selection of appropriate minimization technologies;
- Selection of construction materials with minimal impurities (including materials used for modifications and improvements);
- Implementing appropriate operational procedures in terms of radioactive discharges and managing radioactive waste;
- Good practice awareness and training programmes for personnel.

2.1.1. Minimization of radioactive waste volumes

2.1.1.1. Radioactive waste classification

The IAEA Safety Guide GSG-1 in Ref. [4] describes a radioactive waste classification system for different waste classes (Fig. 1). This classification system for different waste classes (Fig. 1) is based on activity, radionuclides half-lifes and the related minimum acceptable degree of containment and isolation for safe disposal [4]. As part of the regulatory framework, each Member State need to establish a waste clasification system that can be based on waste properties such as radiological, physical, and chemical. Note that for the classification of radioactive hazardous waste the presence of hazardous constituents properties (e.g., chemical toxicity, biohazards) can be more restrictive than the radiological aspects when defining a radioactive hazardous waste classification system.

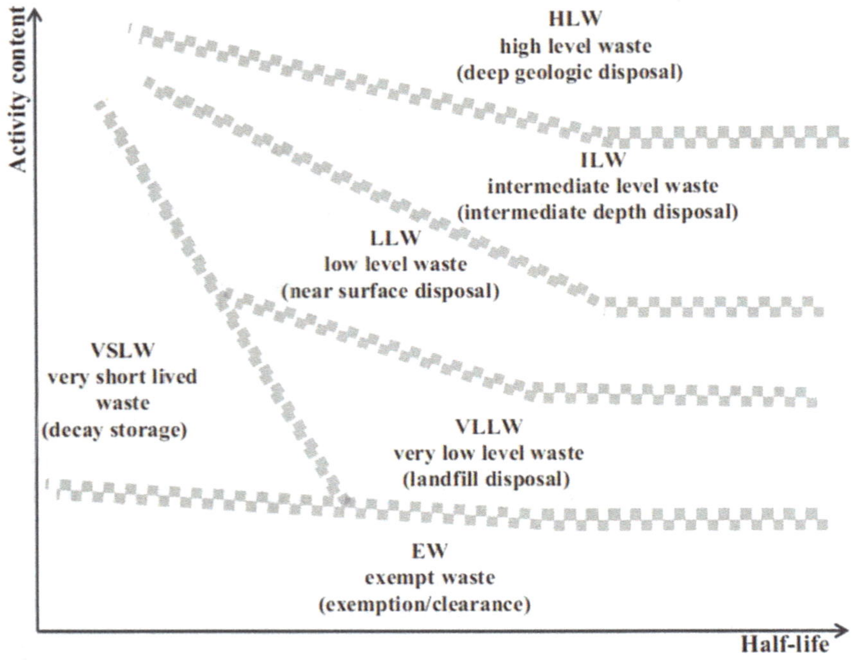

FIG. 1. IAEA waste classification based on the activity of the waste. (adapted from [4]).

2.1.1.2. Solid waste

Minimizing the volume of solid radioactive waste is possible by:

- Reducing or preventing the spread of radioactivity that may generate secondary radioactive waste;
- Establishing appropriate waste collection points and segregation areas before processing;
- Selecting construction material to prevent waste arises through irradiation;
- Selecting facility consumables;
- Establishing and applying clearance level policies;
- Reviewing procedures and policies in a regular manner.

A readily achievable approach to reduce waste volumes is by implementing technical and administrative measures to prevent or minimize the spread of radioactive contamination. Implementation of such measures will reduce the need for decontamination, thereby minimizing associated secondary waste generated by decontamination activities. The extent to which measures are put in place to limit contamination spread can be defined within economically justifiable boundaries. Procedures, appropriate tools, and trained staff can help to identify, confine and remove contamination in a timely manner to minimize the spread of radioactive contamination during operational activities.

Ideally, appropriate collection and waste segregation areas, for further processing, storing or release, need to be considered in the design stage of a facility. In addition, these areas should be able to be adapted over the course of operations in case of changes to the programme such as a change in radiological protection requirements.

Selecting metallic construction materials with minimum impurities that could, through neutron activation, result in gamma emitting radioisotopes (e.g., ^{60}Co, ^{108m}Ag and ^{110m}Ag) could contribute to a volume reduction (minimization) of activated metal arisings. For instance, where concrete is used as a biological shield surrounding the reactor vessel, trace elements (Cs, Co, rare earths) present in the manufacturing admixtures, as well as in the reinforcing metal, needs to be limited to reduce the neutron activation [5].

Adequate minimization strategies could also involve selecting facility consumables in a way that facilitates easier processing or reduces secondary waste for disposal including the use of cellulose, rubber, plastics, lead, filters, thermal insulation, grit blast mediums, and wood during NPP operations, as is further discussed in Section 6.

Developing and implementing clearance procedures for solid waste materials will facilitate an increase in the quantities of solid waste that can be released from regulatory control, thus reducing the volume of active solid waste that needs to be processed.

Management process, particularly waste minimization related policies (e.g., strategy, plan/programme, procedures), needs regular assessment to determine the effectiveness of waste related provisions. Sharing practical and proven examples of implemented waste minimization procedures among NPP operators will result in an overall improvement of waste minimization in the industry.

2.1.1.3. Liquid discharges

Reducing the volume of liquid waste can be achieved in several ways. One example is by performing waterless decontamination activity. Another example is to discharge a certain amount of liquid from the facility, within the approved envelope of licensed liquid release limits, on a periodic basis reducing the volume to be processed.

When applying for authorized discharge, numerous factors (economic and social) need to be considered. Therefore, it will involve dialogue between the regulator, the operator and interested parties, particularly those who might be affected by the discharges (local communities). The IAEA Safety Standard GSR Part 3 [6] (Requirement 8) states: "The regulatory body shall approve which sources, including materials and objects, within notified practices or authorized practices may be cleared from regulatory control", and IAEA Safety Standard GSG–9 [7] gives more detailed guidance on how to develop discharge authorization and discharge limits. Regulators establish the discharge limits based on annual dose constraints (as a part of the annual dose legal limit for population). For example, the United States of America Secretary of Energy requested a 50 percent reduction in operational waste (based on 1993 baseline) for NPPs operated under the Department of Energy (DOE) and DOE has reported that it has achieved this guideline [8].

The activity and volume of radioactive discharges/effluents released from individual NPPs are tightly regulated, and the actual discharged effluents (both in terms of volume and activity concentration) tend to be lower than the limits set or approved by the regulator, as a consequence of the applied measures and improvements in minimizing radioactive effluents. For example, at the site of the Diablo Canyon NPP, there has been a notable decrease of radioisotope activity concentration in the released wastewater during the last two decades as illustrated in Fig. 2.

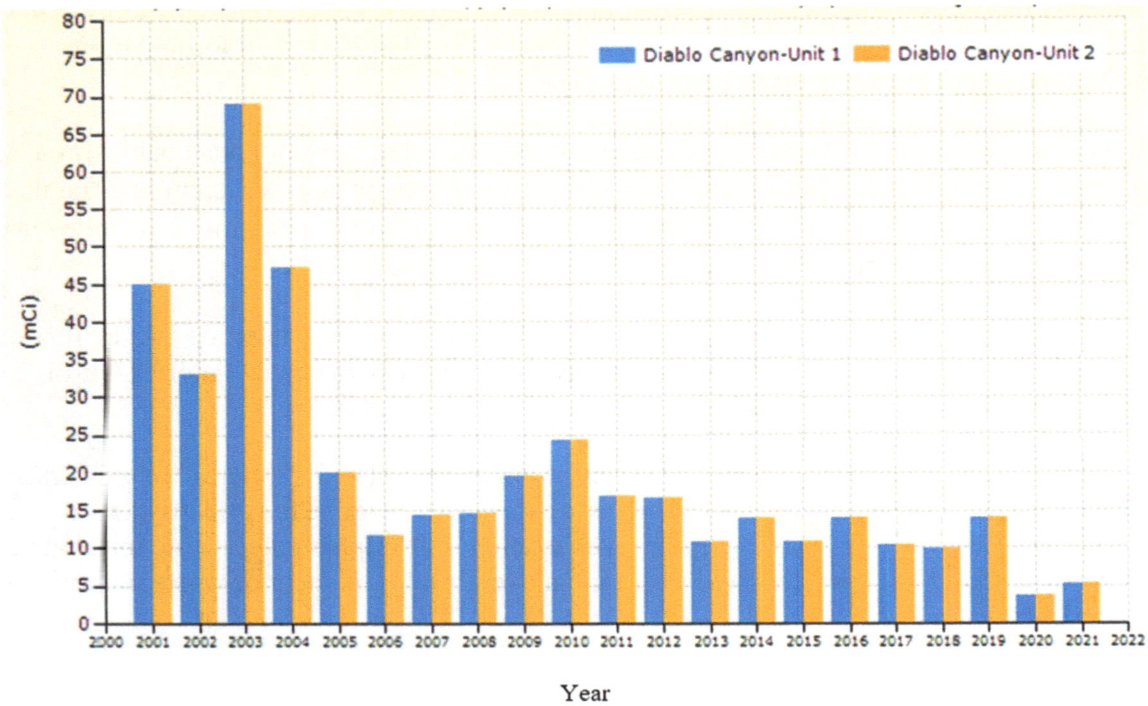

FIG. 2. Total activity of radioactive effluents at the Diablo Canyon Power Plant. (Courtesy of Pacific Gas and Electric Company, USA).

Goals and strategies for dealing with radioactive effluents have also been influenced by international agreements, such as the Convention for the Protection of the Marine Environment of the North-East Atlantic (OSPAR) [9], Protection of the Marine Environment of the Baltic Sea Area (Helsinki) [10], Barcelona Convention [11], Convention on Environmental Impact Assessment in a Transboundary Context (Espoo) Convention [12] and Kyiv Protocol [13].

2.1.2. Recycling and reuse of materials

An important element of waste minimization is to recycle and reuse materials from potential waste streams. Planning in advance for the recycling or reusing of materials and equipment, buildings, and sites, will significantly reduce disposal volumes.

The reuse and recycling programme within Électricité de France (EDF) Group consists of decontamination and remelting of contaminated metals (as illustrated in Fig. 3).

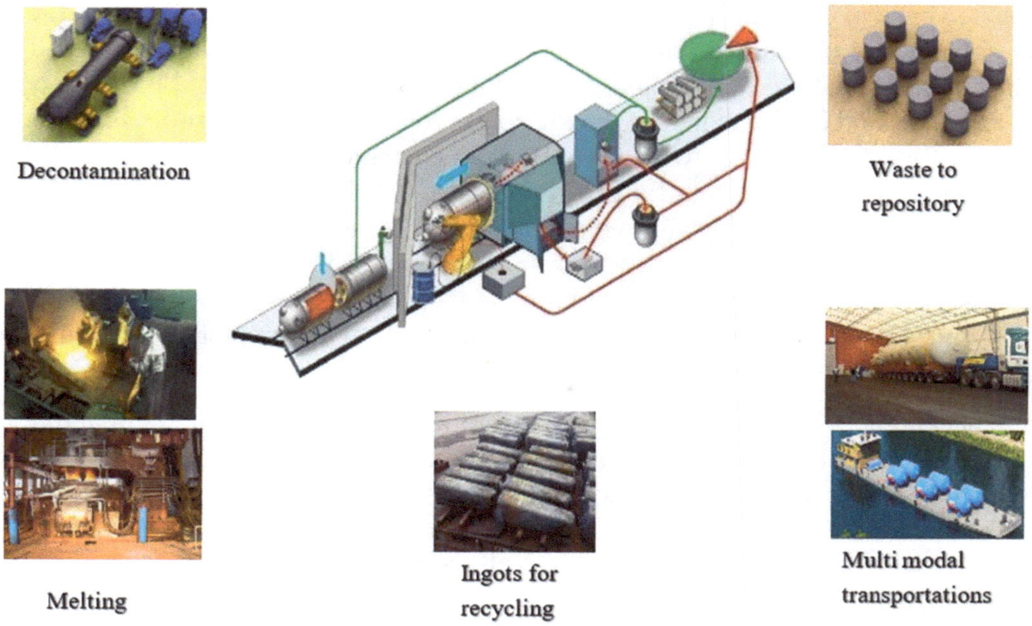

FIG.3. Waste Minimization Strategies: Current Applications of Reuse and Recycling Principles within EDF Group (Courtesy of EDF, France).

In a recycle and reuse management plan, suitable criteria, measurement methodology, instrumentation, work planning and a preventative maintenance programme needs to be specified. Once the recycle and reuse management plan has been defined, proper work instructions need to be developed and implemented with the aim to recycle or reuse of equipment, instrumentation, and materials.

2.2. WASTE MINIMIZATION OBJECTIVES

The objectives of a waste minimization programme could be influenced by:

— Fee structures for disposal: The cost for processing and disposal of large volumes of waste will provide good incentives for minimizing the amount (volume) of waste for disposal. A disposal fee based on the volume of waste to be disposed will encourage waste minimization practices. Alternatively, disposal facilities could also have a combination of 'fixed-plus-volume-based' fees. In this type of arrangement, the baseline fixed fee covers construction and operating costs of the facility, and the additional variable volume-based fee (based on cubic meter of volume used for disposal) will encourage waste minimization practices as well;

— Internal waste processing costs: A volume-based fee paid by the government agency, contractor and/or, NPPs to a processing organization will aid purchasing new or advanced volume reduction technologies to support waste minimization;

— Storage capacity limitations: Limited storage capacity and restrictions against expanded capacity (without jeopardizing safety aspects) will support waste volume minimization. For example, NPPs periodically reviewing the volume of waste authorized for storage each year and reducing the authorization limit annually (NPP operator or regulator) will encourage waste minimization [14];
The National Nuclear Safety Administration (NNSA) of the USA (national programme) is proposing that annual waste generation goal of less than 50 m^3 per each 1000MW unit [15]. In China (national programme), the dry active waste (DAW) volume during outages is

targeted to be between 20 and 35 m^3 for the China National Nuclear Power Co., Ltd (CNNP) while for Sanmen (a sub plant of CNNP), the objective to be achieved is ≤20 m^3 [16];

— Performance Indicators: Performance indicators between NNPs could promote consistent, high standards and establish industry wide continuous improvement processes to reduce the amounts of waste arises. Currently, the World Association of Nuclear Operators (WANO) publish performance indicators for low level waste (LLW) arises from NPPs [17];
— Financial compensation: By allowing contracted companies to specify the amount of protective clothing or supplies and consumables needed for a project on the NPP site, unnecessary waste could be generated. Financial incentives, based on waste volumes (arises) minimization, needs to be negotiated not only with contacted companies but also with NPP staff. This is common practice in the USA [18];
— Public acceptance: The value of a positive public perception that the radioactive waste is managed responsibly by keeping the amount to a minimum and implementing new activities to reduce it further is well recognized [19].

2.3. POLICIES INFLUENCING MINIMIZATION STRATEGIES

Policies will influence the selection of waste minimization practices, and equally, the economic incentives (cost of minimization activities and disposal costs) can influence policies.

2.3.1. National regulations and policies

National regulations and policies can either promote or discourage waste minimization practices. The IAEA publication [20] is identifying activities to be considered during the design of nuclear facilities, including NPPs to minimize waste generation. The National Nuclear Safety Administration (NNSA) of the USA has published a national guidance of "Waste Minimization for Nuclear Facilities" applicable to USA NNPs [15], that confirms the additionally need for a safety analysis report (FSAR), which require to include waste minimization principles, requirements, organization, responsibilities. and provides guidance on the content of waste minimization annual reports and record keeping.

In the absence of a national policy promoting waste minimization practices over a NPP life cycle, NPP operators need to initiate dialogue with legislators and competent authorities to establish a waste minimization policy.

2.3.2. Safety policies

Safety requirements and guidance to facilitate the establishment of safety policies for radioactive waste management (including waste volume reduction) are set out and discussed in detail in various IAEA Safety Reports and Safety Standards Series publications. These set out the principles of radioactive waste management safety for various scenarios and some of the most useful are:

— General Safety Requirements Part 3, Radiation Protection and Safety of Radiation Sources [6];
— Safety Standards Series No. GSG–9, Regulatory Control of Radioactive Discharges to the Environment [7];
— General Safety Guide No. GSG–16, Leadership, Management and Culture for Safety in Radioactive Waste Management [21];

- General Safety Requirements GSR Part 4 (Rev. 1), Safety Assessment for Facilities and Activities [22];
- Specific Safety Guide SSG–40, Predisposal Management of Radioactive Waste from Nuclear Power Plants and Research Reactors [23];
- Specific Safety Guide SSG–41, Predisposal Management of Radioactive Waste from Nuclear Fuel Cycle Facilities [24];
- Safety Guide WS–G–6.1, Storage of Radioactive Waste [25];
- Safety Guide RS–G–1.7, Application of the Concepts of Exclusion, Exemption and Clearance [26];
- General Safety Guide GSG–1, Classification of Radioactive Waste [4];
- General safety guide No. SSR–2/1 (Rev1), Safety of Nuclear Power Plants: Design [27];
- General safety guide No. NS–G–2.2, Operational Limits and Conditions and Operating Procedures for Nuclear Power Plants [28];
- General safety guide No. NS–G–1.13, Radiation Protection Aspects in the Design of Nuclear Power Plants [29];
- General safety guide No. SSR–2/2 (Rev. 1), Safety of Nuclear Power Plants: Commissioning and Operation [30];
- General safety guide No. GSG–7, Occupational Radiation Protection [31];
- Safety Standard Series SSG–52, Design of the Reactor Core for Nuclear Power Plants [32].
- Safety aspects need to be fully considered during the development and implementation of a waste minimization plan.
- Administrative policies and procedures
- Administrative policies, controls, and procedures as well as management initiatives in operating facilities can contribute significantly to an optimal waste volume minimization strategy.
- Management structure and assignment of responsibilities
- The operational management of NPPs can contribute significantly to the establishment of an adequate waste minimization strategy by initiating an organizational structure which lists the responsible groups for each aspect of the waste volume minimization programme as waste arises in different departments.
- Management needs to ensure that proper staffing is available, and training is provided in order to implement waste volume reduction techniques and practices in the various departments.
- Alternatively, an independent waste management organization could be designated to be responsible for the waste minimization programme of a NPP thereby increasing visibility and accountability within the organization. As examples:
- The EXELON generation group in United States of America (USA) employs a Corporate Functional Area Manager (CFAM) standard peer group for the application of the waste minimization programme [33];
- In Japan, the Alara review board (consisting of different CFAM) is responsible for the waste minimization programme for Tokyo Electric Power Company (TEPCO) [34];
- In France, EDF (85 reactors worldwide) created EDF-DP2D and its associated holding, Cyclife, which both gather all skills and industrial capacities to address waste management for the whole EDF group (170 000 personnel) [35];
- In Italy, Societa Gestione Impianti Nucleari S.p.A (Sogin) is responsible for waste management for all NPP facilities [36];
- In Spain, Empresa Nacional de Residuos Radiactivos S.A. (Enresa) is responsible for the waste management of NPP facilities [37].

2.3.3. Ownership of a waste management programme

For a successful implementation of a waste minimization programme, management needs to take ownership and use it as a NPP performance indicator to ensure continuous improvement. The waste minimization programme needs to be a focused document with a single, knowledgeable programme owner, relying on designated radioactive waste support representatives from other key sections (operations, maintenance, environment, chemistry, radiation protection, quality assurance, and engineering supports).

Ownership of management programme includes facilitating awareness by educational programmes and training employees in managing waste. Such education programme could embed considerations in the fields of contamination control, quality control of materials, proper specification of product, process modification and selection of applicable tools and equipment.

2.3.4. Waste minimization plan

Waste minimization plans can be established at the level of each NPP operational section/unit if the applicable waste minimization programmes are established and enforceable. They will differ, but the following could be considered to be included into a waste minimization plan:

— Source reduction measures;
— Leak reduction measures;
— Periodic cost analysis of processing and disposal costs;
— Accounting and tracing system to quantify the waste arisings;
— Types, amounts, activities and dispositions of contaminated materials;
— Improvement of operational techniques;
— Educational or training programme;
— Near and long-term initiatives;
— Actions defined for owners with due dates;
— Implementation of innovative techniques with proven increased efficiency;
— Status report to senior management (annually updated);
— Graphics reporting a minimalization progress (annually prepared).

The waste minimization plans also need to consider factors such as disposal routes for contaminated material, quantities of material arisings, worker doses, costs for storage of wastes and costs saving from recovered materials.

A technical example of minimizing secondary waste is shown in Figure 4 whereby polyethylene sheeting was set up during maintenance activities to reduce secondary waste.

FIG. 4. Reduction in secondary waste by selection of polyethylene covering material (Courtesy of Ontario Power Generation, Canada).

2.3.5. Training plan

Major volume reductions in waste arisings can be achieved when the personnel understand the need for minimizing waste during the execution of their duties, and they are performing these duties in full compliance with the operational procedures. This is achievable by providing technical and safety training courses and confirming learning materials with visual aids (e.g., videos, photos) illustrating similar activities [38]. Training has to fully adhere to the as low as reasonably achievable (ALARA) principles and could include the following (depending on individual roles):

— Identification of waste arisings and the respective minimization techniques;
— Segregation guidelines;
— Monitoring procedures;
— Safety and health programme;
— Direct ('hands-on') training;
— Duties, tasks and responsibilities;
— Reporting structure;
— Use of knowledge management information;
— Training using representative mock-ups.

2.3.6. Consideration during design

Technical factors to consider during the design phase to ensure minimization of waste arisings during the NPP lifetime are [20]:

— Design of installations (including the choice of the construction site);
— Materials selection considering activation, chemical interaction and corrosion;
— Maintenance planned for designed installations.

2.4.5.1. Design of installations

Building layout and component installation need to comply with safety requirements. However, where possible, pipes and components need to be designed for easy drainage and accessibility. In a liquid system, the number of flanges, joints, elbows, and crude traps need to be reduced to prevent crud (solids) build-up that could result in an increase in activity, leaks, and widespread contamination. Welded rather than flanged joints can reduce leakage of contaminated liquids but could impede removal difficulties during dismantling.

2.4.5.2. Materials selection

The properties of material selected could influence the maintenance schedule and frequent replacement of the defective components will result in solid waste volume increases and thus, to ensure a minimum maintenance, the materials with the following attributes are suggested:

— Impurities with low activation properties;
— Resistance to high temperatures;
— Corrosion resistant;
— Long term structural integrity;
— Chemical stable;
— Machinable (repairable).

Metal selection for the primary coolant system or metals/materials in the vicinity of the reactor core need to contain minimum impurities since, through neutron activation, long-lived and/ or strong gamma emitting radioisotopes could be produced. For instance, if the use of alloys containing nickel cannot be avoided, the alloy with lowest nickel content needs to be considered.

Currently, metals/materials/alloys with minimal cobalt content are being developed to minimize the neutron activation of stable ^{59}Co that lead to form radioactive ^{60}Co which is one of the major radionuclides of concern during operating NPPs. The metal selection requirement (minimal cobalt content) not only applies during the design of a NNP but also during the decades of operation while considering the replacement of components or retrofitting of systems.

Processing (removal, treatment, disposal) of contaminated surfaces as waste can be minimized by materials that can be decontaminated to clearance levels (more easily than concrete flooring) thus reducing the volume of waste. For instance, the use of an epoxy or similar coating or even steel flooring that can be decontaminated to unconditional clearance levels for re-use could be considered. (For more information see Section 6).

2.4.5.3. Maintenance of designed installations

Adequate and routine maintenance of materials throughout a NNP plant will minimize replacement of defective operational materials. In general, cranes, electrical transformers, ventilation systems and instruments need to be properly inspected and maintained (especially parts that may degrade) and replacements need to be carefully selected to ensure a reduction in waste volume for disposal.

2.4.5.4. Process optimization

There are avenues for optimizing a waste minimization process, and this can be achieved through:

- Staff feedback – established programmes need to include feedback from the staff as this could be incorporated into the corrective action programme;
- Periodic assessments – periodic evaluations could identify deficiencies resulting in defining corrective actions;
- Industrial operational experience – periodic search for information from operational facilities to identify opportunities for improvement. Those searches may identify actions to prevent equipment failure, and could by implementation minimize negative feedback from periodic evaluations;
- Periodic safety evaluations – these may be incorporated into the assessments of the integrated management system;
- Industrial benchmark – active research possible improvement opportunities and lessons learned via remote communications, forums, or facility visits for process optimization;
- Interaction with vendors and suppliers – direct interaction with equipment providers and subject experts to evaluate current performance and identify process enhancements;
- Review of technical publications – The Institute of Nuclear Power Operations (INPO), industrial organizations (e.g., World Association of Nuclear Operators (WANO), Electric Power Research Institute (EPRI) and World Nuclear Association (WNA), are producing technical reference publications that support process optimization and are based on actual experience and/or technology development;
- The IAEA guideline on Development and Implementation of a Process Based Management System [39] - provides practical guidance to nuclear organizations that are planning to implement a management system to comply with IAEA Safety Standards Series No. GSR-Part 2 [40]. These two Safety Standards are worth to be consulted by newcomer countries,

since a vendor-provided "management system" delivered with a nuclear power plant to ensure safe operation is often a quality management system for operations and maintenance, which may integrate aspects related to safety and environmental protection;

— The IAEA publication on Invitation and Evaluation of Bids for Nuclear Power Plants [41] emphasizes the integrity and interdependence of various activities related to invitation of bids, their technical and economic evaluation, and contracting. It provides the necessary information to organize, guide and realize these activities as an integrated process. Furthermore, this publication indicates how and to what degree the activities preceding the preparation of the technical specification, evaluation of bids and contracting can influence the process.

2.4. WORK OPTIMIZATION

Work optimization could facilitate the success of a waste minimization programme between different departments in a NPP facility [42]. Examples of work optimization (planning, processes, innovation) will be described in this Section.

2.4.1. Work planning

Work optimization implies planning the activities before execution, for instance :

— Periodic safety evaluations/reviews – these may be incorporated into the assessments;
— Work process (planning cycle) and software;
— Multi-discipline team planning;
— Worker training and qualification;
— Clearly defined waste disposition plans by activity embedded in work control process;
— Optimization of radiation protection plans (radiation risk estimation versus benefit);
— Use of scans;
— 3D imaging, CAD drawings that support virtual visuals of components/work areas;
— Training maintenance work in inactive circumstances for increasing work efficiency;
— Increasing material storage capacity for providing time to decay activity content;
— Better work planning, eliminate need for entry into radiological areas.

2.4.2. Parameters to be considered in work optimization

Work optimization needs to consider a broad range of radiological and non-radiological detriments to optimize work execution by including:

— Work process (planning cycle) and software;
— Health risks from toxic chemicals;
— Ionization exposure;
— Accidents;
— Environmental impacts;
— Economic impacts.

2.4.3. Innovation and use of robotics

Another incentivization for optimization could be the application of innovative processes, resulting from research and development or new international practices being shared among operators thanks to international cooperation.

In that frame, the increasing use of automated means and robotics appears as a means of waste minimization and waste management optimization [43]. Indeed, the emerging field of robotics in nuclear operation during the lifetime of a NPP, allows for the progressive automation of common tasks like maintenance (Fig. 5). Generalizing the use of remote handling tools or robots for traditional nuclear human operation (maintenance, design modifications) permits reduction in:

— The volume of secondary waste (e.g., clothing and air locks);
— The number of packages (with the bigger/heavier parts being segmented and carried out);
— Labour by mitigating job preparation/termination periods (donning and doffing protective clothing, PPE usage, obtaining personal dosimeters, activation of work permits),

while increasing the:

— Safety by minimizing the risk of injury to operators;
— Disassembly efficiency (less cuttings, enabling bigger/heavier parts being cut and evacuated).

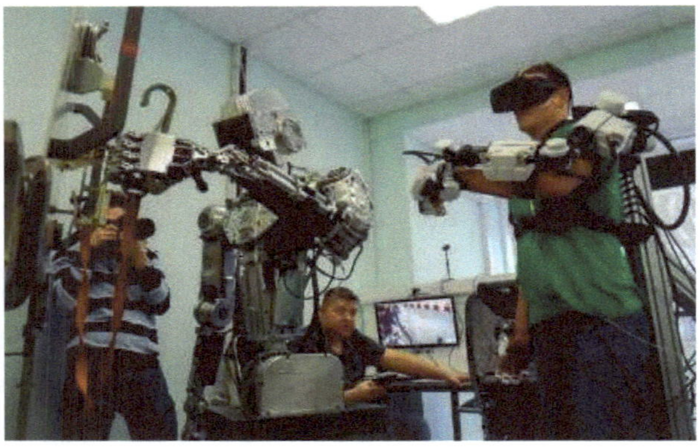

FIG. 5. EDF Industrial demonstrator facility for robotic development. (Courtesy of EDF, France)

3. LIQUID WASTE MINIMIZATION STRATEGIES

3.1. LIQUID WASTE ARISINGS

Potential liquid effluent volumes at a NPP are influenced by reactor type, operating methodology and facility design while the activity levels are a result of:

- Metal type and impurities in reactor vessel metal;
- Material composition of components in contact with coolant;
- Controlling the chemistry of reactor coolant;
- The quality of fuel cladding material;
- Fuel cladding integrity;
- Contamination on cladding during manufacturing.

In NNPs, the main source of liquid radioactive waste can be attributed to the primary coolant and water from the fuel storage pools. Other liquid radioactive waste at a NPP, as well as the methodologies to minimize the volume of liquid waste streams, are summarized in Table 1.

3.2. MINIMIZING AQUOUS LIQUID WASTE GENERATION

3.2.1. Programmatic controls

Successful liquid waste minimization management (volume reduction) requires programmatic controls such as:

- Establishment of an effective leakage control and prioritization programme for the primary coolant system and other connected systems;
- Use of the work planning and execution process to ensure equipment and systems are maintained in their design configuration;
- Define drain paths and segregation schemes for all waste liquids arising during plant operating and maintenance evolutions;
- Implementation of surveillance and testing / sampling protocols to monitor system performance;
- Establishment of programmes to monitor the secondary liquid arisings (closed cooling, service water, fire protection and chilling water);
- Planning of liquid waste treatment methods (removing of radionuclides and process handling);
- Retention and reuse of the boric acid solutions drained from the primary loop systems during outage;
- Separation of housekeeping solutions (e.g., mop water and decontamination solutions) from non-active and radioactive effluents;
- Verification of alkalization during treatment of liquid effluent to avoid the solidification and reduce secondary waste arisings.

TABLE 1. LIQUID WASTE TYPES, ORIGIN AND REDUCTION TECHNIQUES

Waste Type	Waste origin	Reduction practices
All liquids excluding organic liquids	Auxiliary system leaks	Reduce leaks of service water
	Primary to secondary system leaks	Reduce primary leaks
	Reactor coolant system leaks	Reduce leaks from pumps and valve seals
	Steam generator blow-down purification	Release of inactive regeneration solutions
	Draining of boric acid solutions during outages	Increase quality of procedures and operation manipulations
	Regeneration solutions from ion exchange purification	Optimize the periodicity and duration of regeneration
	Drain water purification	Regeneration (distillation)
	Regeneration of ion-exchange resins	Improve the quality of procedures and operational manipulations
	Decontamination	Optimize/Adopt the concentration of decontamination solutions
	Laundry Processes	Optimize the usage protective clothing and PPE. Control of the contamination of the personnel.
	Storage pool lining leaks	Increase quality of repairs
	Showers	Control the contamination of the personnel. Collect the shower water separately from water streams with higher contamination.
	Collection tanks	Perform monitoring and discharge the low activity water
	Laboratories and activities	Optimize sampling periodicity and sampling quantities. Optimize calibrations or use of software calibration.
Solvents	Cleaning operations	Regeneration (distillation)
	Decontamination solutions	Regeneration (distillation)
Oil	Leaking Equipment	Repair leaks
Resin and filter cartridges	Liquid purification system	Use more efficient resins Use non-metal cartridge filters
Sludges and slurries	Collection tanks Overflowing tanks Sedimentation tanks Piping	Protect input into technological system

Examples of methodologies at NPPs to minimize the volume of aqueous liquid waste streams (primary coolant, fuel storage pool and radiologically controlled area (RCA) of the NPP), as well as examples of methodologies to minimize the volumes of other types of liquid waste (e.g., organic liquid and decontamination liquids) will be discussed in the following sections.

3.2.2. Segregate collection of liquids

Any non-radioactive water-based liquid generated in the radiologically controlled area (RCA) of the plant could either be routed outside the RCA or collected in separate sumps or tanks within the RCA for monitoring and discharge without treatment. This will reduce the volume of liquid waste that needs to be processed.

Liquids with low activity levels need to be collected in separate sumps or tanks for monitoring and discharge without treatment after a short storage period (allowing the decay of radionuclides).

Furthermore, segregated collection of radioactive high purity liquids (e.g., equipment drains) from radioactive low purity liquids (e.g., floor drains) needs to be considered.

During outages the separation of borated liquid waste streams from non-borated wase streams could reduce the volume of drained liquids waste (borated liquid).

Both new water-water energetic reactors (WWER) and western light-water reactors (LWR) are designed with a segregated collection option of waste liquids. Older NPP operating plants that were not designed with segregated collection of liquids systems, could implement modifications to reduce liquid volumes to be further conditioned. The flow chart (Fig. 6) shows the segregated collection tanks for the liquid radwaste system at Diablo Canyon NPP.

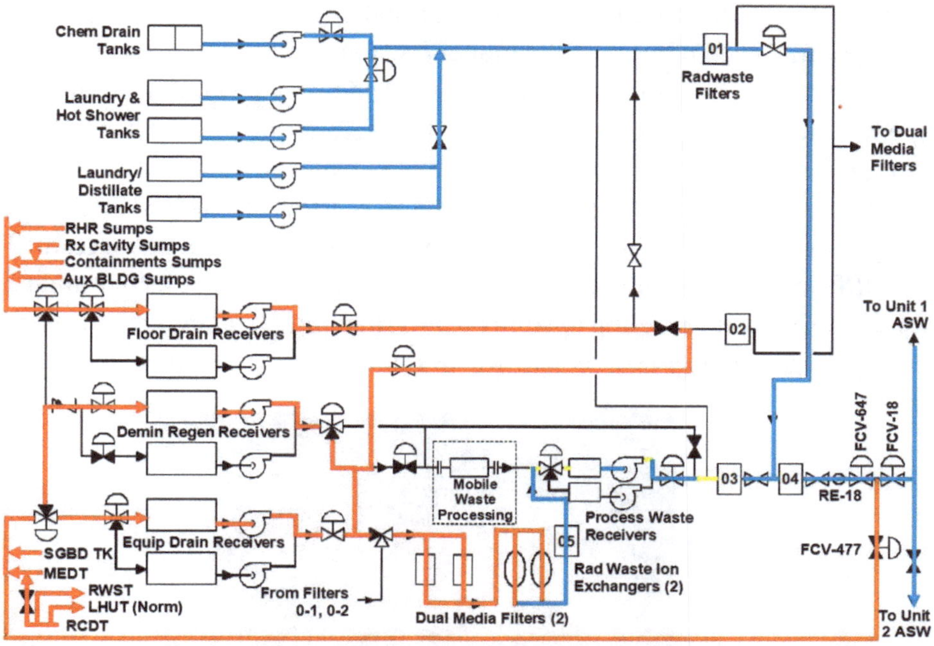

FIG. 6. Segregated collection tanks for the liquid radwaste system at Diablo Canyon Power Plant. (Courtesy of Pacific Gas & Electric Company, USA).

At Diablo Canyon NPP, separate collection tanks are provided for laundry drains, decontamination shower drains, and laboratory drains. All these liquids are low in radioactivity, monitored and discharged through filters to the Pacific Ocean (as indicated by the blue lines). Separate tanks are provided to collect equipment drains and floor drains. These liquids are sent through the process system as indicated by the red lines to the processed waste receiver tanks. The treated liquid is then monitored and discharged.

In Brazil at the Angra 2 NPP, the liquid waste arisings are segregated into three groups (collection tanks) based on the origin, contamination, and processing for instance as:

— Group I – KPK11-12 BB001 (70 m^3): Medium level liquid waste and liquid wastes with boric acid contents and specific activity in the range of 3,7E+06 to 3,7E+08 Bq/m^3 as well as liquid waste from drains of rooms which contain radioactive components, laboratories, and decontamination facilities.
— Group II – KPK13-14 BB001 (70 m^3): Inactive to low level waste and specific activity in the range between 3,7E+04 and 3,7E+06 Bq/m^3 as well as liquid waste from building sumps of service rooms and regenerate water from steam's generator blowdown system.
— Group III – KPK15 BB001 (70 m^3): Wastes with detergent contents (e.g., laundry and showers) in the controlled area.

The storage of liquid waste (Group III) in tank KPK15 BB001 (70 m^3), resulted in a saving of approximately 140 m^3/month of liquid waste to be evaporated, minimizing evaporator concentrate for immobilizing into bitumen.

3.2.3. Reuse and recycling of liquids

The reuse of liquids is an excellent radioactive waste reduction strategy. For instance, the reuse of RCA treated liquid (purified to reactor coolant requirements) not only reduces radioactive processing costs but also the volume of additional makeup water. When incorporated into the initial plant design, the size and cost of the makeup water system can be reduced. Additionally, treated liquid from the RCA, with specifications outside reactor coolant requirements can be reused for spent resin transfer operations within the RCA of the plant. Using RCA water from a reused water tank, instead of new clean demineralized water, will minimize the liquid arisings. Normally clean demineralized water or RCA water is used to flush systems (contaminated tanks, receiving tanks and pipe) after transferring boron concentrate or spent resin. The example from Angra 2 NPP (in Fig. 7) indicates the system used for directing clean drainage water from the RCA into the boric acid recycling systems using flushing connections, hoses and collectors thereby reducing the amount of radioactive waste liquids during outages.

FIG. 7. Reuse of water from RCA drains during outages to minimize waste at the Angra 2 NPP. (Courtesy of Electrobras Electronuclear, Brazil).

Other examples that can be considered to reduce liquid volumes are the recycling of liquids containing chemicals, for instance the reuse of boric acid liquids and the recycling of decontamination solutions. The usage of filters or sorbents for particulate (with adsorbed radionuclides) could facilitate the re-use of the liquids.

3.2.4. Leak repair

The philosophy of eliminating all leaks and placing a high priority on the repair of new leaks is a key minimization practice. Radioactive liquid from leaks in the RCA are often accumulated in sumps with contents being pumped to liquid waste storage tank(s). Decontamination liquids to clean spillages are also pumped to liquid waste storage tank(s) thereby increasing the volume of accumulated liquid waste in these tanks. The use of maintenance teams with pre-approved work orders for equipment repairs leaks have aided in the success of volume reduction of liquids in the storage tanks. The following processes could be introduced as goals in a minimization plan:

— Reduce the number of physical leaks within the RCA initially to zero leaks over a set time period;
— After a refuelling outage, ensure no (zero) physical leaks in the RCA (plant condition objective linking to a waste minimization objective);
— Repair leaks that arise during operation within a set time period (e.g., three months). Drip bags could be considered (Fig. 8) for leaks;
— Implement temporary modifications to reduce/re-route leakages;
— Establish daily surveillance programmes for the potential valves, pumps and tanks to minimize the leakage;
— Prevent groundwater and/or precipitation intrusion to reduce the quantity of liquid system influents.

FIG. 8. Drip bags used for leaks (when allowed by regulator). (Courtesy of Pacific Gas & Electric Company, USA).

As radioactive liquid from leaks in the RCA accumulated in waste storage tank(s), it is difficult to determine new leaks by monitoring the level of the tank. Similar issues exist for measuring volume contents in the different individual sumps inside the RCA. Angra 2 NPP has developed a software that monitor all tanks regarding pumping volumes and liquid levels to immediately identify anomalies that could indicate new leakage. Monitoring can be also done by operators in the field or remotely through instrumentation.

3.2.5. Other activities

Installation of metal filters on floor drains to constrain solid impurities from going inside the liquid waste collection system could be considered (Fig. 9).

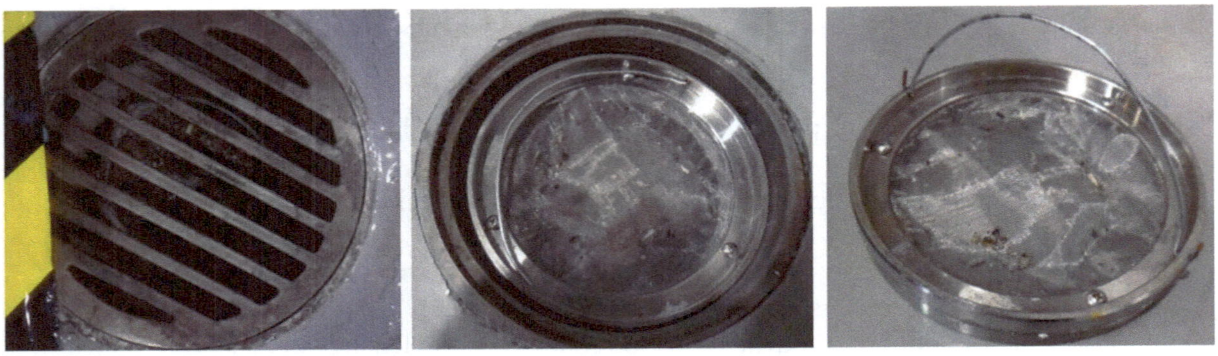

FIG. 9. Installing a metal filter on floor drain to remove solids. (Courtesy of Pacific Gas & Electric Company, USA).

Mop water can either be dispositioned with laundry drains or collected in separate containers (Fig. 10) and dried in a drum dryer (Fig. 11) with the drums dispositioned as radioactive waste.

FIG. 10. Contaminated Mop Water in drums and saddle tank. (Courtesy of Pacific Gas & Electric Company, USA).

FIG. 11. Drum Dryer for Mop Water to Prevent Fouling Floor Drains. (Courtesy of Pacific Gas & Electric Company, USA).

Another option is to consider routing radioactive reverse osmosis (RO) liquid back to the collection tanks for reprocessing through the upstream ion exchange vessels. The activity in this RO reject liquid often converts into a precipitate that can be filtered. By cleaning of waste holdup tanks to remove accumulated solid precipitates, the number of cartridge filters will be consistently reduced. (Fig. 12).

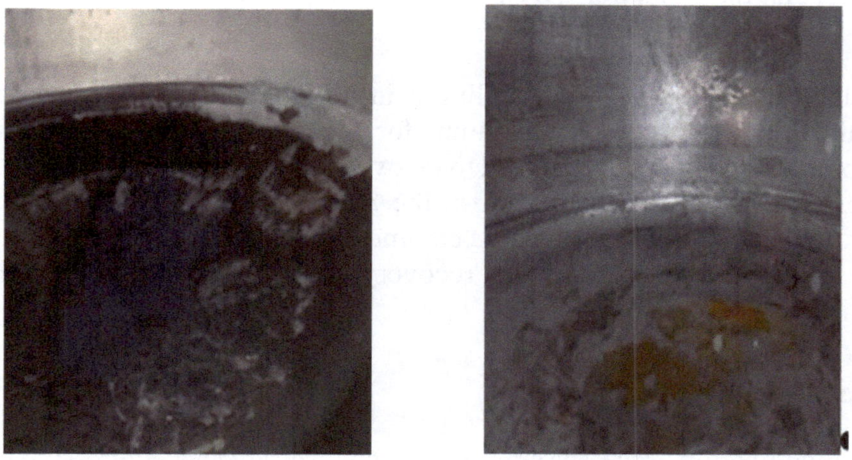

a) *Before* b) *After*

FIG. 12. Cleaning of waste holdup tanks to remove solid precipitates. (Courtesy of Pacific Gas & Electric Company, USA).

The use of dryer technology for evaporator or RO concentrates to remove water will avoid the increase in waste volume from cement solidification. For plants with both dryer and cementation technologies, drying high activity concentrate may generate a higher waste class package, while cementation of low activity concentration waste may result in an unnecessary volume increase.

3.3. MINIMIZING ORGANIC LIQUID WASTE

Oil arises from NPP operations are representing a small volume of the total of liquid waste arisings. Recycling of radioactive oil for reuse is possible by:

— Removing contaminants by re-extraction;
— Filtering (polishing) of oil to remove radioactive contaminants.

For example, synthetic oil is used in the four reactor coolant pumps at Diablo Canyon Power NPP. During a refuelling outage, the oil from one of the four pumps is replaced and the used oil is collected in drums. The used oil is then 'polished' using a small portable pump and filter system. Samples of the polished oil are sent to an off-site laboratory for analysis and if within specifications, the recovered oil (polished oil) is reintroduced into a reactor coolant pump during the next refuelling outage.

3.4. MINIMIZING DECONTAMINATION LIQUID

There is no universal decontamination process, and the decontamination methods need to be selected based on effectiveness, compatibility with the system or component, secondary waste generation and waste disposal, as well as radiological and industrial safety [44].

Liquid waste arising from decontamination activities, could contain different complexing chemicals that needs to be processed for disposal. Processing is difficult as different decontamination solutions with complexing chemicals requires different technologies leading to individual waste matrixes for disposal (volume increase for disposal). To standardize the processing requirements for specific complexing chemicals, at Krško NPP (Slovenia) was decided to standardize the commercial products throughout the NPP, for instance only to allow for use Duschgel BB (special decontamination shampoo with additive disinfectant from Hakawerk), Dekontafix (paste for decontamination) and Deko soap (liquid soap for washing hands) [45].

For the decontamination of uranium spillages in NPP facilities, different decontamination solutions could be used to complex uranium for removal. As uranium is under safeguards control, the complex needs to be processed to recover the uranium. In this regard the conversion facility and the SAFARI research reactor at the South African Nuclear Energy Corporation (Necsa, South Africa) standardized a single commercial decontamination solution to be used to limit the use of technology of uranium recovery from these organic complexes found in decontamination solutions.

3.4.1. Strippable gels technology

Chemical decontamination processes are based on concentrated and aggressive liquid solutions. During this process, the volumes of liquid radioactive waste that are generated needs to be further treated. Société des Techniques en Milieu Ionisant (STMI) has been collaborating with the French Centre National de la Recherche Scientifique (CNRS) laboratory to develop processes using gels and foams (waterless processes) to decontaminate stainless steels, alloys, and painted surfaces [46]. High decontamination factors exceeding 2000 can be obtained, while the volume of waste arisings is minimal when compared to the use of water-based solutions. Advantages of gels and foams as decontamination mediums to replace liquid based technologies include:

- No need for preparatory activities;
- Straightforward application (apply → dry → peel → dispose);
- Penetrates cracks, pores, and voids;
- Applicable on different surfaces;
- Respirators not required during application;
- Does not generate volumes of liquid waste to be processed, only compatible solid waste.

Examples of different gel processes used at NPPs, which replaced water-based technologies and resulted in only compactible, low-volume solid secondary waste, are presented in the followings.

3.4.1.1. DeconGel

DeconGel (a commercial product) is a neutral-pH, low-odour polymer hydrogel (comprising a combination of surfactants, chelating agents, thixotropic agents, wetting agents, defoamers, biocides, and buffers) used for the decontamination of concrete, steel, painted or unpainted surfaces, glove boxes, hoods, and casks without the use of water [47]. Once dried, the product traps the contaminants in a polymer matrix, and the resulting film can be peeled off and disposed, according to appropriate regulations (Fig. 13).

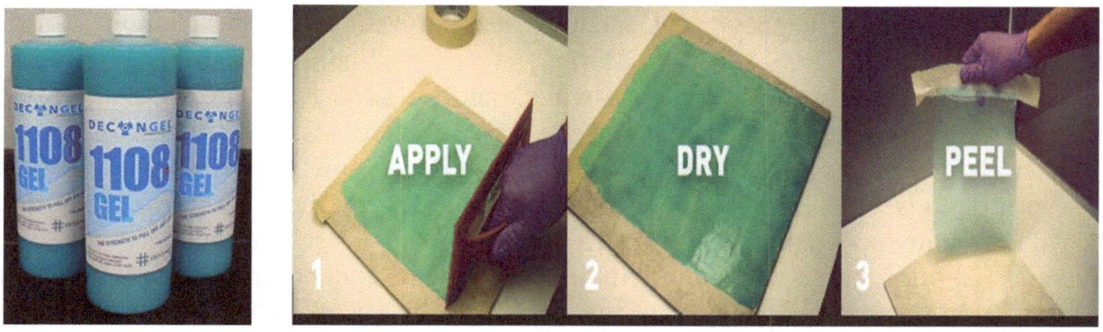

FIG. 13. Application of DeconGel to remove contamination.

DeconGel can be used to:

- Decontaminate the adsorbed radioisotopes on metals: ^{3}H, ^{14}C, ^{18}F, ^{125}I, ^{131}I, ^{99}Tc, ^{201}Ti, ^{241}Am, ^{238}Pu, ^{239}Pu, ^{240}Pu, ^{241}Pu, ^{242}Pu, ^{137}Cs, ^{134}Cs, ^{60}Co, ^{54}Mn, ^{55}Fe, ^{63}Ni, ^{59}Ni, ^{90}Sr and ^{58}Co;
- Remove contaminants from toxic industrial materials: asbestos, polychlorinated biphenyls (PCBs) and oil;
- Remove the chemical spills.

Different formulations of commercial products of DeconGel are currently available:

- DeconGel 1102 – used for the removal of petroleum-based contaminants, including PCBs and oil spills;
- DeconGel 1108 – used for toxic industrial chemicals (TIC) and toxic industrial material (TIM);
- DeconGel 1128 – Spray-on version of 1108 to cover large areas.

DeconGel is used at NPPs for decontamination, for instance at Perry, Ontario Power Generation, Bruce Power, Peach Bottom, Constellation (Byron, Braidwood, Nine Mile Point). DeconGel is used to decontaminate of concrete, painted diamond metal plates, steam generators (prepared for disposal) and leaks and spills. Fig. 14 showcases floor decontamination at a Canadian NPP following a heat exchanger leak [47].

a) Before *b) After*

FIG.14. Application of DeconGel to remove contamination after a leak at a Canadian NPP (reproduced from Ref [47] with permission courtesy of CBI Polymers, Canada).

The decontamination efficiency of DeconGel was investigated for cleaning a uranium conversion facility at NECSA (South Africa), which included contaminated equipment, polypropylene floor coverings, concrete walls, and pressed roofs. Results indicate that decontamination efficiency decreases as surface porosity increases. The findings are presented as a case study in Annex I. In Slovakia, the nuclear and decommissioning company [company name] Plc applied DeconGel to the inner surfaces of the tanks at the V1 NPP site (building CB 801A: V1) to remove contamination (Fig. 15). Once dried, the gel was peeled off, collected in commercial Tatrafol bags, and stored in drums for subsequent processing. The level of decontamination was verified using abrasion tests.

a) Before *b) After*

FIG. 15. Decontamination of tanks at Bohunice V1 NPP during decommissioning. (Courtesy of JAVIS, Slovakia).

The welds of the stainless-steel sheets forming the body of the tank were identified as problematic. The welds contained pores with 226-Ra contamination, requiring more aggressive decontamination methods. Additional details are available in Annex I.

The spraying of decontamination gel on building surfaces at the Fukushima Daiichi NPP to bind, encapsulate, and remove radioactive contaminants is illustrated in Fig. 16 [48].

FIG. 16. Application of gel to remove contamination at Fukushima Daiichi NPP (reproduced from Ref [48] with permission courtesy of Tokyo Electric Power Company (TEPCO)).

Other examples regarding the use of gels (excluding DeconGel) are:

— A polyvinyl alcohol–borax complex containing adsorbents for the decontamination of ^{137}Cs-contaminated surfaces [49];
— A nitric-acid-loaded polymer hydrogel with a high removal of ^{137}Cs and ^{90}Sr on stainless steel [50];
— A strippable hydrogel composed of polyvinyl alcohol and phenylboronic acid grafted alginate for the decontamination of ^{137}Cs contaminated surfaces [51];
— A strippable coating using acrylate emulsion as the main film-forming agent and lauryl sodium sulphate as surfactant for removing or uranium dust on the concrete surface [52];
— Chitosan gels with or without Fe_3O_4 nanoparticles for removal of uranium contamination [53].

3.4.1.2. SuperGel

SuperGel, a commercially available product developed at Argonne National Laboratory, contains engineered nanoparticles within a polymer gel to decontaminate surfaces. It was designed as a fast-acting emergency response to a 'dirty bomb' or other radioactive dispersal contamination [54]. The decontamination efficiency depends on the number of applications, the specific radioisotopes involved (e.g., strontium, cobalt, americium, plutonium, uranium), and the characteristics of the contaminated structure. SuperGel was later reformulated to include a mild additive that promotes the dissolution of metal scale, such as rust [55].

The application of SuperGel follows a simple, three-step process: application, reaction, and clean-up, as shown in Fig. 17. The gel is applied to an area (left) and removed after three hours (middle). The final activity of the cleaned area (right) was measured using HPGe and alpha/beta radiation detectors, confirming that it was suitable for disposal.

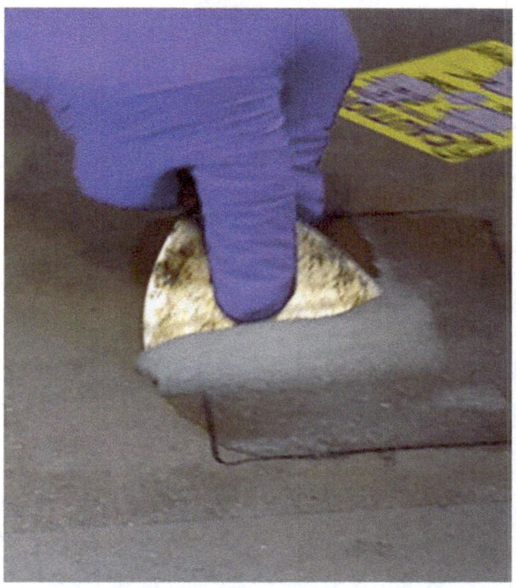

FIG. 17. SuperGel application to a contaminated surface (reproduced from Ref [55] with permission courtesy of CEA, France).

3.4.2. Vacuumable gels (powders)

Vacuumable gels, also known as 'self-drying gels,' cracking gels (SDC Gels), or by the commercial name ASPIGEL, contain mineral colloidal particles dispersed in a gel solution that encapsulate radioactive contaminants through sorption. As the gel dries, it fractures into a solid powdery that can be removed by vacuuming (Fig. 18).

a) Application *(b) Drying (cracking)* *(c) Final powdery solid*

FIG. 18. Example of decontamination with gel: (reproduced from Ref [56] with permission courtesy of CEA, France).

Different commercial self-drying or cracking gel (with the different compositions) for the decontamination of metallic surfaces (steel, stainless steel, aluminium, and lead) are currently available:

— ASPIGEL 050 or 052: Dryable gels for the decontamination of 'fatty' materials. Not compatible with aluminium alloys. ASPIGEL 052 does not contain organics;
— ASPIGEL 100 or 100V: Dryable gels for the decontamination of stainless steel;
— ASPIGEL 100E: Dryable gels for the decontamination of stainless steel, does not contain ammonium ions;
— ASPIGEL 200: Dryable gels for the decontamination of ferrous metals;
— ASPIGEL 300: Dryable gels for the decontamination of lead;
— ASPIGEL 400: Dryable gels for the decontamination of aluminium alloys;
— ASPIGEL 500: Dryable gels for the decontamination of concrete.

Self-drying or cracking gels are applied by spraying onto stainless-steel surfaces, as shown in Fig. 19 [56]. The gel is sprayed using specific nozzles to ensure homogeneous coverage of the contaminated surface. Once dried, the resulting flakes containing the contamination can be collected for disposal.

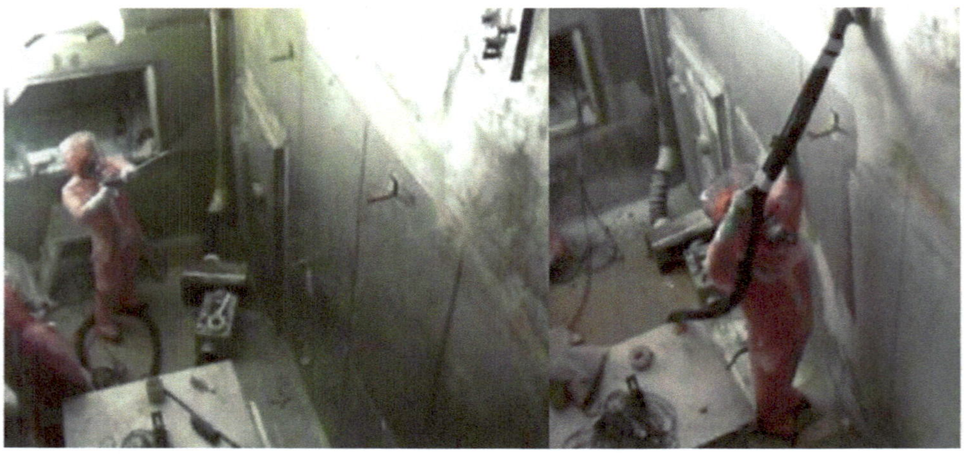

FIG. 19. Spraying of a self-drying gel onto a contaminated area and the removal of cracked gel after drying (reproduced from Ref [56] with permission courtesy of CEA, France).

3.4.3. Foams

For decontamination of complex internal configurations (e.g., inside of pipes), foam decontamination is preferred (Fig 20). Decontamination foam contains chemicals and wetting agents (containing hydrophilic or hydrophobic groups) capable of penetrating the pores of the components, effectively decontaminating all the exposed surfaces [57]. Additionally, foam recirculation can enhance the decontamination process, ensuring maximum efficacy.

FIG. 20. Using foam technology for pipes decontamination (reproduced from Ref [56] with permission courtesy of CEA, France).

For maximum efficacy, it is particularly important to optimise the following when considering using foams as decontamination solutions:

- The compatibility/durability of the decontamination components in foam with the target material;
- Foam stability during the decontamination process;
- Removal of foam from the object that need to be decontaminated;
- Processing and handling the secondary waste containing residual organic compounds (surfactant and eventual organic thickener) and contaminants.

France conducted a study on the use of foam technology for decontaminating austenitic steel heat exchangers, specifically to dissolve chromium and remove contamination [58]. The foam formulation included Ce (III)/Ce (IV) as redox agents, along with biodegradable surfactants. The results indicated that the stability of the foam could be maintained for several days. After the decontamination process, the resulting liquid, known as collapsed foam, was filtered and pumped into interim storage tanks. To achieve a near-neutral pH fluid (pH 6), alkaline or acidic liquids were mixed into the decontamination solution before further processing, which included the coprecipitation of radionuclides.

4. WET SOLID WASTE MINIMIZATION STRATEGIES

4.1. WET SOLID WASTE ARISES

At NPPs, potential wet solid waste arises in the form of filtration media, concentrates and sludges. Although wet wastes are low in volume, they have high activities and require quite complex treatment and/or conditioning prior to disposal. As such, wet waste processing is typically expensive to disposition, and minimization of waste volumes is considered.

The objective of wet solid waste minimization programme is to reduce the volume, the mass and the activity of the radioactive waste at the source thereby reducing the use of purification methods (e.g., filtration media) that increase the volume of wet solids (e.g., sludge) requiring conditioning.

The nuclear industry is continually improving efforts to reduce wet solid waste volumes, as shown in Fig. 21 (yearly trend) for the Diablo Canyon NPP.

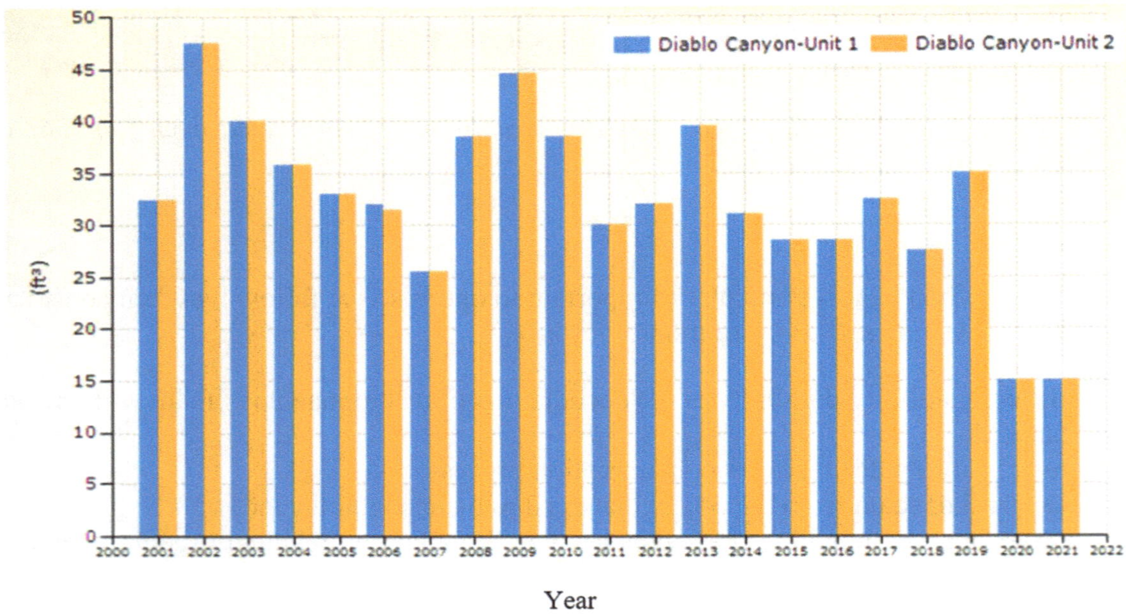

FIG. 21. Yearly Wet Solid low and Intermediate Level Waste Arises at Diablo Canyon NPP. (Courtesy of Pacific Gas & Electric Company, USA).

An example of wet solid waste is sludge, which typically consists of a combination of iron, magnesium, calcium and aluminium hydroxides. However, if chromium is present, the risk associated with processing increases due to its toxicity. Therefore, minimizing sludge waste volumes will not only reduce disposal volumes but also lower processing risks.

Another source of wet solid waste is filter media (such as ion exchange resin, activated carbon and cartridge filters). When exhausted (saturated), these filters are removed from service and treated as radioactive waste.

The use of ion exchange media contributes to this waste streams, as the ion exchange process exhibits the following limitations:

— Non-selective ion exchange (IX) resins have an extremely low capacity for effluents with high inactive salt content (>1 g/l);
— A high concentration of suspended solids in the liquid effluent to be treated may decrease efficiency due to blockage;
— When gamma-emitting radionuclides are removed, the radiation dose can become significantly high, requiring a very heavy lead shielding configuration;
— Organic ion exchange resins are prone to both thermal and radiation degradation;
— Certain chemical conditions can also lead to degradation.

Wet waste in the form of evaporation concentrates contain a mixture of boric acid, fission and activation products. The industry is continually improving performance and waste minimization, as shown in Fig. 22 for concentrate generation in Slovenia.

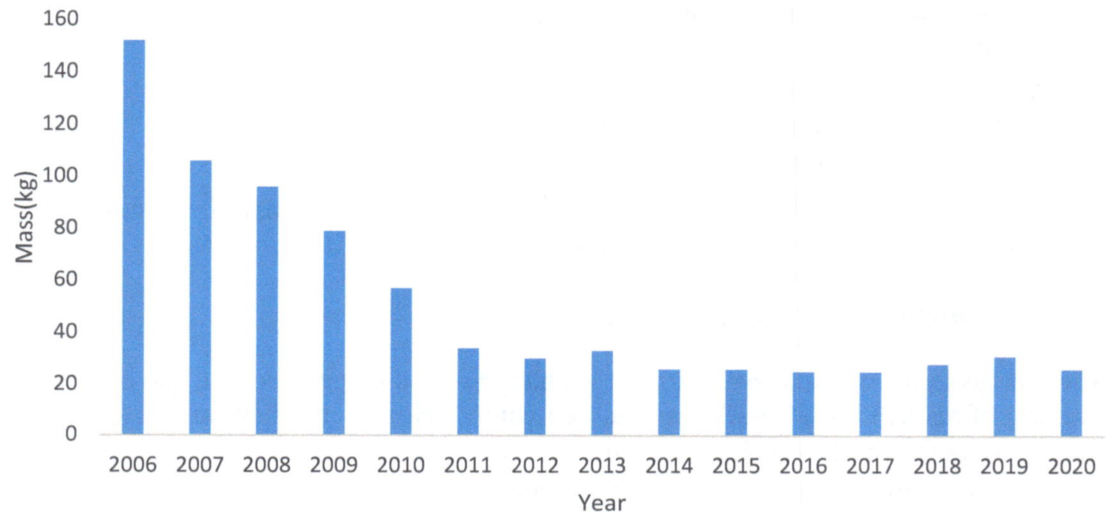

FIG. 22. Year average of concentrate at KRŠKO NPP. (Courtesy of Slovenske Elektrarne, Slovenia).

4.2. PRACTICES TO REDUCE WET SOLID WASTE VOLUMES

By controlling the chemical composition of the primary coolant, moderator systems and reactor auxiliary systems, the formation of corrosion products and precipitates, wet solid waste volume (such as sludge) formation and the use of purification requirements (e.g., filtration media) can be minimized. A well implemented chemistry programme is not only essential for the safe operation of a NPP but also crucial for reducing wet solid waste volumes. Therefore, the following considerations are important:

— Safety of plant structures, systems and components is as intended by design;
— Chemical impurities and the build-up of activated corrosion products are minimized, thereby reducing radiation exposure;
— Low levels of radioactivity and proper chemistry control are maintained in the system water through the purification system, which uses filters and ion exchange resins;
— The circuits associated with these systems contain sources of radioactivity and are therefore the main contributors to radioactive waste and personnel exposure;

— The inventory of activated corrosion and erosion products in the systems is minimized by:
  - Chemistry controls to reduce corrosion;
  - Equipment design features that limit erosion;
  - Proper selection of material with low activation products (e.g., using low cobalt impurity materials)
— The inventory of fission products in the system is minimized by:
  - Managing fuel defects through proper fuel design, in-core fuel management and chemistry control;
  - Removing elevated activities using filters and ion exchange resins.
— A proactive chemistry monitoring programme that evaluates whether control parameters are within a technically specified range.

The technologies for the removal of corrosion products and precipitate formation are as follows:

— Ion exchange resin and other absorbing media;
— Media filtration;
— Cartridge filters;
— Recycling of boric acid concentrates to minimize arisings;
— Control of foreign material exclusion.

To reduce the volume of wet solid waste requiring processing, some operational guidelines are applied and are briefly presented in the followings.

4.2.1. Ion exchange resins and other absorbing media

In purification systems where a series of ion exchange resins is required, the use of mixed bed resin (mixing of resins in one vessel) need to be avoided. However, where only a single vessel is available to perform the purification, the ration of the cation/anion resins need to be investigated to maximize the life of the mixed bed. The goal is to ensure that both types of resins become fully loaded at nearly the same time. Typically, standard stochiometric mixed beds (two parts anion to one part cation) has to be avoided, as a ratio of two parts cation to one part anion is preferred. This is because recent improvements in chemical control of water systems have reduced the demand for anion resin. In purification systems where ion exchange resins are housed in separate vessels with individual media, the volume is better regulated. For example, when a specific resin becomes fully loaded with impurities, only the loaded resin needs to be replaced, rather than the entire system. This approach reduces the volume of wet solids requiring processing. The use of advanced resin types is worth investigating, as these resins are more than three times more resistant to radiation, reducing the need for replacement due to the breakdown of their organic structure.

Inorganic ion exchange media, such as zeolite, has a much higher capacity for caesium isotopes than cation resin. Using a zeolite bed upstream of the resin vessel can extend the life of the cation resin, thereby decreasing the overall volume of wet waste. The use of selective ion exchange media might also be considered for the removal of difficult-to-remove nuclides (e.g., $^{124/125}Sb$, ^{63}Ni).

For spent resin with low activity, dewatering the media for storage or disposal can reduce the volume of waste, provided that regulations and disposal criteria allow. For high activity resin, consider dewatering in a high integrity container to further reduce the waste volume, if permitted.

Organic ion exchange materials can be regenerated using strong acids or bases (to remove absorbed radionuclides) for reuse, but this process generates radioactive liquid waste with a high salt content that requires proper disposal. Fully absorbed spent ion exchange resin needs to be stabilized, and polymerization may be considered to mitigate storage impacts, as shown in Fig. 23 (additional information available as case study in Annex II.)

FIG. 23. Volume reduction of spent ion exchange resins before processing - Mercure process (Courtesy of EDF, France).

4.2.2. Media filtration

Activated carbon (charcoal) filtration media is used in NPPs to remove oil, organics, and particulate matter upstream of resin beds, preventing blockages in the resins. This reduces the need for resin replacement and minimizes the additional volume of wet waste. For operational use, Sanmen Nuclear Power Co. Ltd in China washes the granular charcoal with demineralized water to remove charcoal debris that could block filter cartridges during operation, which would otherwise increase the volume of wet solid waste [16]. Activated carbon (charcoal) is washed between three to five times, until all carbon fines are removed (Fig. 24). Although artificial zeolite may generate fewer fines than natural zeolite, also requires washing.

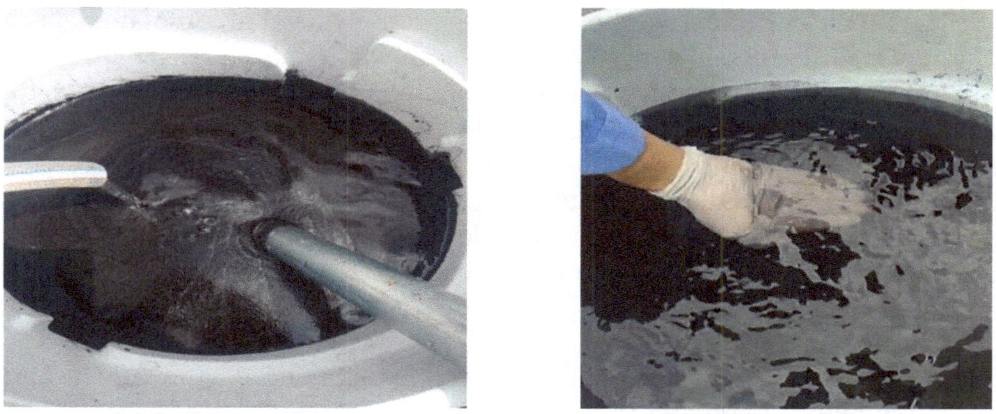

(a) before washing *(b) after washing*

FIG. 24. Removal of carbon debris that could block filter cartridges during operation (courtesy of Sanmen Nuclear Power Co. Ltd., People Republic of China).

4.2.3. Cartridge filters

Cartridge filters are probably the most common used filtration technology for particulate removal, serving as a pre-treatment step for downstream ion exchange and membrane technologies. They are also used as a pre-treatment step for evaporators and reactor purification systems.

Advanced planning and careful evaluation of available cartridge filter options are necessary to ensure the appropriate filter is selected to meet the plant's filtration needs, while minimizing unnecessary waste generation. The following considerations can be included in the planning and evaluation process:

— Define factors impacting spend filter generation (e.g., influent water quality challenges);
— Define service objectives (e.g., water clarity);
— Define micron ratings for various operational modes or activities – lower micron ratings are not always better and may increase the frequency of filter replacements. It is important to select the correct filter pore size for the intended service while minimizing radioactive waste generation;
— Test filters – before use, conduct a tabletop or laboratory test to ensure the filter performs as expected and to assess the distribution of activity in the spent filter.

Once in service, monitor dose rate and activity build-up to ensure filters do not exceed pre-established radiological limits. If necessary, consider shearing or shredding the spent filters to increase waste packaging efficiency. A disposal volume reduction (VR) ratio of 4:1 can be achieved through this process. If waste packages containing filters exceed pre-established radiological limits for low-level waste (LLW) disposal, the intermediate level waste (ILW) surface dose rate classification could be reduced by encapsulating the spent filters in cement. This encapsulation helps lower the dose rate on the surface of the waste container, enabling compliance with disposal site requirements.

4.2.4. Recycling of boric acid concentrates to minimize waste arisings

Boric acid is used in pressurized water reactors (PWRs) as a soluble neutron absorber, capturing neutrons that would otherwise contribute to the nuclear fission reaction. This enhances shielding, control, and safety during the most critical stages in nuclear energy operation. During the NNP operation, metal impurities form corrosion products and radionuclides (^{60}Co, ^{137}Cs, ^{90}Sr and ^{3}H) within the boric acid solution. When the activity level becomes high, the boric acid needs to be replaced. Processing (purifying) of boric acid from NPPs for reuse reduces the volume of boric acid that require processing. Processing of boric acid concentrates for reuse in NPPs was introduced at Loviisa NNP (Finland) and Paks NPP (Hungary) [59]. The process entails the removal of suspended solids from the boric acid solution, followed by a pH adjustment step to facilitate the precipitation of sodium tetraborate. The sodium tetraborate precipitate is then removed using pressure filtration. The recovered solid sodium tetraborate is subsequently reused by dissolving it into the primary water circuit to the achieve the required specified boric acid concentration.

At Paks NNP, the process was further modified by removing or decontaminating ^{60}Co and ^{137}Cs from the solution before removing the suspended solids through physico-chemical methods. The resulting sodium tetraborate, free of 60Co and 137Cs, can either be reused at the NPP or disposed of as inactive hazardous waste.

4.2.5. Control of foreign material exclusion

The presence of 'foreign' materials, such as debris and fines from the fuel assemblies in the water pool, can act as radionuclide absorbers but also as catalyst for sludge formation. A stringent control policy for foreign material (especially during refuelling outages) can help limit the use of additional resin/filters to remove 'foreign' materials and prevent the generation of sludge components. Since implementing such policies at the C-1 facility at Chashma Nuclear Power Generating Station (CNPGS–Pakistan), the volume of spent resin has decreased significantly (Fig. 25) [60].

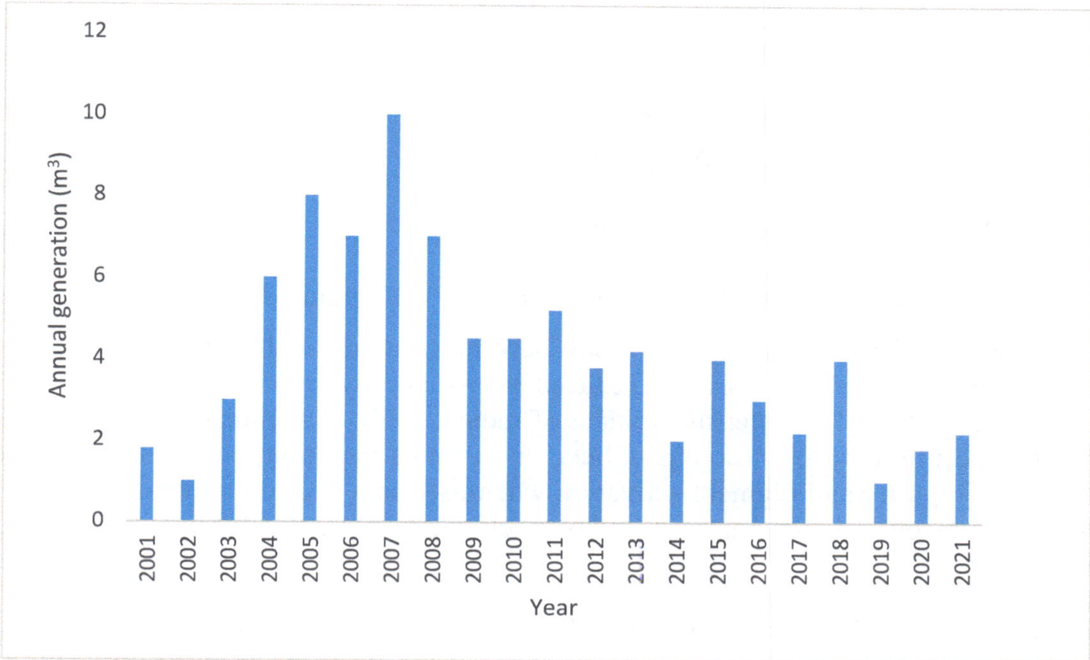

FIG. 25. Annual low and intermediate level spent resin arising from C–1 at CNPGS. (Courtesy of CNPGS, Pakistan).

5. DRY SOLID WASTE MINIMIZATION STRATEGIES

5.1. DRY SOLID WASTE GENERATION

During the normal operational lifetime of a NPP, the composition of dry solid waste varies based on reactor type, operating mode, facility design and other factors. Examples are indicated in Table 2.

The following material types have significant contribution to the waste streams generated by NPPs:

- Cellulose materials (e.g., paper, rags, clothing and wood);
- Rubber (e.g., gloves and boots used as PPE, hoses);
- Plastic (e.g., tubing, sampling vessels, sheeting and traps used to prevent floor contamination);
- Metals (e.g., steel parts of equipment);
- Building debris;
- Lead (e.g., lead blankets, lead sheets and bricks used as shielding and liner material).

Lead blankets and bricks used at NPPs may become contaminated with radionuclides. The contamination of lead blankets can be reduced or avoided by covering the lead plates with plastic sheets, which helps lower the volume of radioactively contaminated hazardous waste. The plastic sheeting is resistant to alpha, beta, and gamma radiation; however, it needs to be evaluated for impurity and pigment activation when used in a neutron radiation environment. Alternatively, uncovered contaminated lead blankets can be separated for treatment, recycling, or disposal.

TABLE 2. DRY SOLID WASTE GENERATION AND MINIMIZATION PRACTICES

	Waste origin	Reduction practices
Clean/Exempted Waste	Equipment from non-radioactive contaminated systems. Decontaminated metals. Filters from the influx ventilation.	Planning and segregate at the point of generation. Measures for prevention of contamination spread to avoid cross- and secondary contamination.
Low Level and Intermediate Level	Parts from the equipment, connected to the primary circuit, water purification systems, radioactive solution pumps and tanks. Ventilation filters	Reuse/recycle programme; Decontamination
Combustible	Routine operation and maintenance	Reuse/recycle programme
Compactable Non-combustible	Routine operation and maintenance	Reuse/recycle programme Change to incinerable materials Segregate and incinerate combustibles Sort for recoverable items
Metal	Routine operation and maintenance	Reuse/recycle programme
Thermal insulation	Routine operation and maintenance	Reuse/recycle programme Separate fiberglass backing from metal sheeting
Filters	Ventilation	Use compressible types Avoid organic solvents Do not replace early
Grit blast media	Decontamination Fragmentation	Reuse/recycle programme Use steel shot or zirconium oxide (last longer
Concrete, rubble, soil	Plant retrofit	Recycled and reuse when possible

5.2. MINIMIZATION OF DRY SOLID WASTE

Reduction of dry solid waste as much as possible is critical for minimizing processing costs and disposal volumes. Examples are shown in Fig. 26.

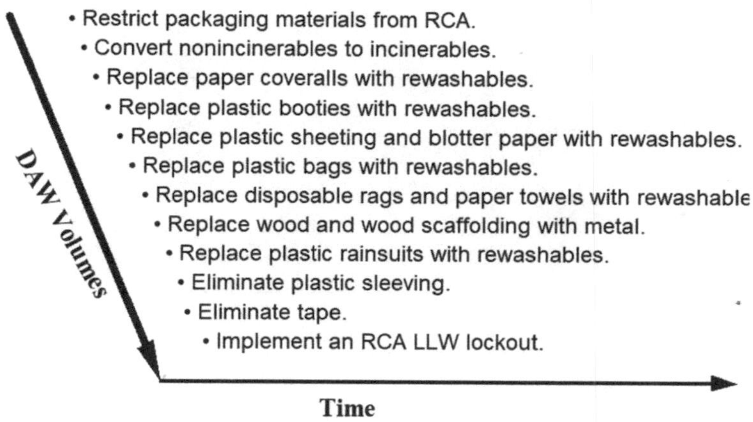

FIG.26. Material control techniques.

Examples of components of a dry solid waste reduction programme include:

- Control of the ingress of packaging materials into the RCA, such as cardboard, plastic and paper;
- Development of an extensive reuse/recycle programme;
- Segregated collection of waste at the source based on radioactive contamination/dose rate and waste types (e.g., metal, combustible, compactable);
- Selection of treatment technologies with high volume reduction factors (e.g., incineration, compaction);
- Preferable use of consumable materials within the RCA that are compatible with the treatment technology;
- Use of modern, easily decontaminated materials for personal protective equipment. The use of additional personal protective equipment (e.g., rubber gloves, boots, respirators) has to be prescribed in accordance with the radiation risk;
- Control of the number of staff entering the RCA;
- Use of water-soluble materials (clothes, rags, foils);
- Restriction of the use of foils for stainless steel floors;
- Inclusion of RAW management issues in the training of employees;
- Provision of pre-job briefing information regarding radiation risk, radioactive contamination, expected RAW generation and RAW handling.

The industry is continually enhancing dry solid waste minimization performance, as indicated by the yearly trend shown in Fig. 27 for the Diablo Canyon NPP.

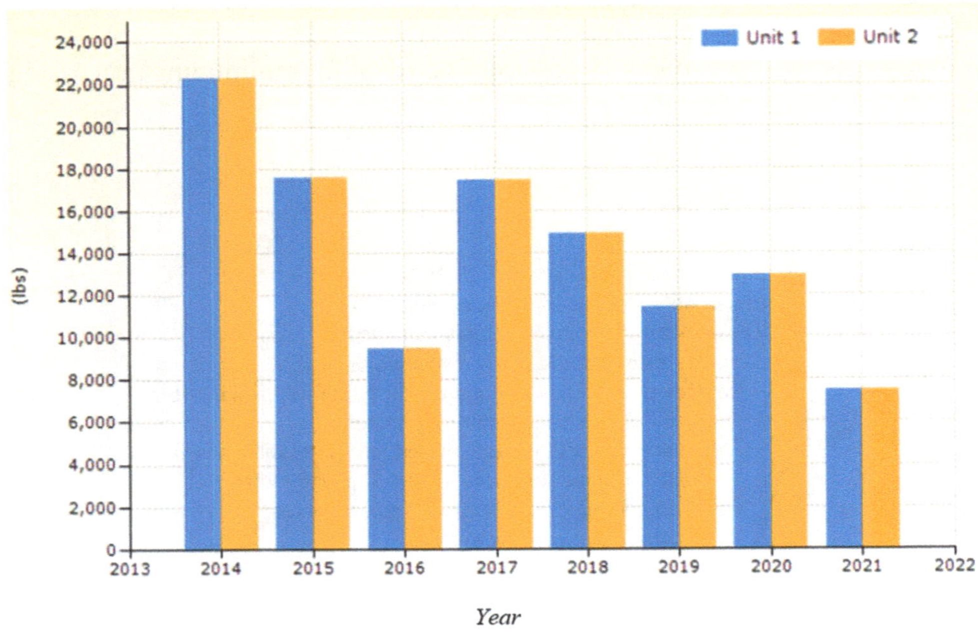

FIG. 27. Yearly average of dry non-metallic solid low and intermediate level waste arises at Diablo Canyon NPP. (Courtesy of Pacific Gas & Electric Company, USA).

5.2.1. Reducing contaminated areas

Reduction of contaminated areas (CAs) within the RCA is crucial for minimizing the volume of dry solid radioactive waste during the decommissioning process. This can be achieved by opting for permanent, easy to decontaminate floor coatings (Fig. 28) instead of using large disposable sheeting that must be disposed of as radioactive waste.

FIG. 28. Floor coatings replacing disposable plastic sheets. (Courtesy of Pacific Gas & Electric Company, USA).

5.2.2. Control of material entering the RCA and worker access control

A programme to prevent unnecessary materials and tools being from being brought into the RCA can be a vital component of any radioactive waste reduction programme. Special focus needs to be given to packaging materials (e.g., cardboard, plastic) common tools, ladders, and carts. Contaminated tool chests or cribs can be reused, especially tools that are difficult to decontaminate.

5.2.3. Considerations about clearance

Pursuing the clearance of wastes generated within the RCA can reduce the volume of solid material that needs to be managed as radioactive waste. Figure 29 illustrates how the application of clearance procedures can significantly reduce the amount of metal waste.

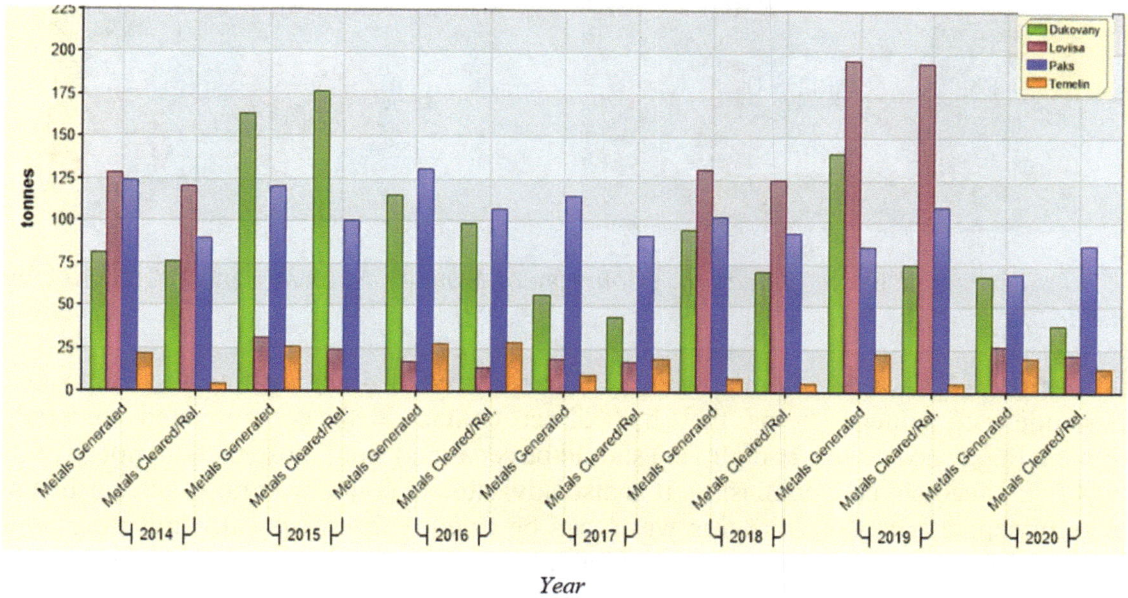

FIG. 29. Generated metal arises vs. cleared metals ((Courtesy of Pacific Gas & Electric Company, USA).

5.3. IMPLEMENTATION OF SEGREGATION AND COLLECTION

For waste generated within the RCA and other nuclear plants in the USA, a national 'Green-is Clean' programme has been implemented. This programme is based on a volumetric assay system for monitoring potentially clean waste, allowing for its unconditionally release and thereby reducing processing volumes. Examples of this national program's implementation can be seen at the Bear Creek Processing Facility [61] and the Oak Ridge National Laboratory [62]. In this programme, containers are provided within the RCA to collect potentially clean waste at the source. Clean items such as chairs, desks, tables, filing cabinets, carts and dollies are surveyed for clearance. Additionally, separate containers for reusable garments can be provided for clearance surveys. Collecting potentially contaminated tools in bags also improves the efficiency of cleaning these tools for return to service.

Another initiative to reduce the volume of operational solid waste (e.g., dissolvable materials, compatibles, metals, wood, power cables for stripping, hoses for shredding) involves the segregations at the source by placing solid materials in different coloured bags. Each NPP station has its own colour designation for bags and containers as illustrated in Fig. 30.

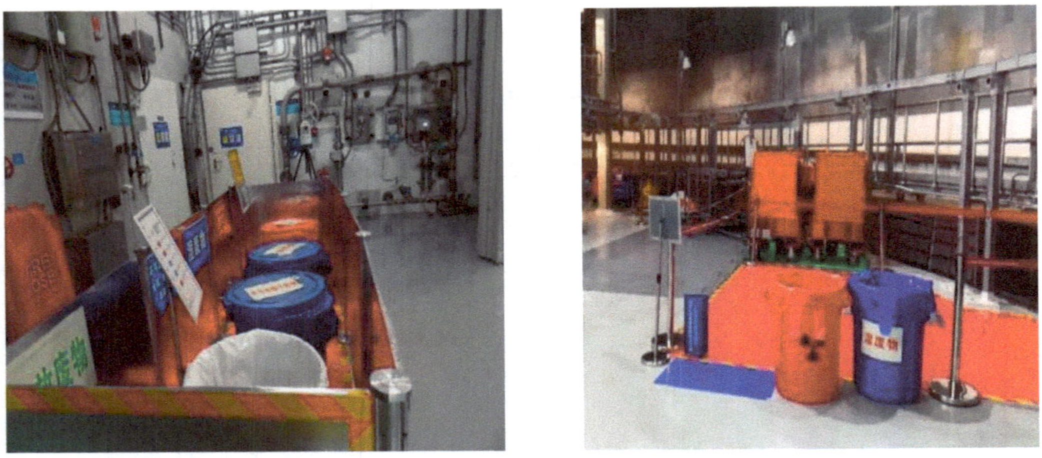

FIG. 30. Segregated waste collection. (Courtesy of Sanmen Nuclear Power Co. Ltd, People Republic of China).

Additionally, solid waste volumes from PPE can be reduced segregating them into distinct colour-coded containers. Items that had direct contact with contaminated materials or equipment (e.g., shoe covers and gloves) should be separated from those that did not have direct contact (e.g., head covers and masks). It is also advisable to collect wet rags, wet garments and oil rags in separate bags so that this waste can be dried before disposal. Collecting metallic waste, typically in metal boxes, not only aids in waste processing but also improves material handling.

For optimum success, staff requires proper training, and adequate containers need to be available in the waste collection and storage areas. Posters or guides (as shown in Figs. 31 and 32) placed near the collection points can help in guiding the staff to place the items in the proper containers.

FIG. 31. Posters near the collection areas facilitating waste segregation. (Courtesy of Ontario Power Generation, Canada).

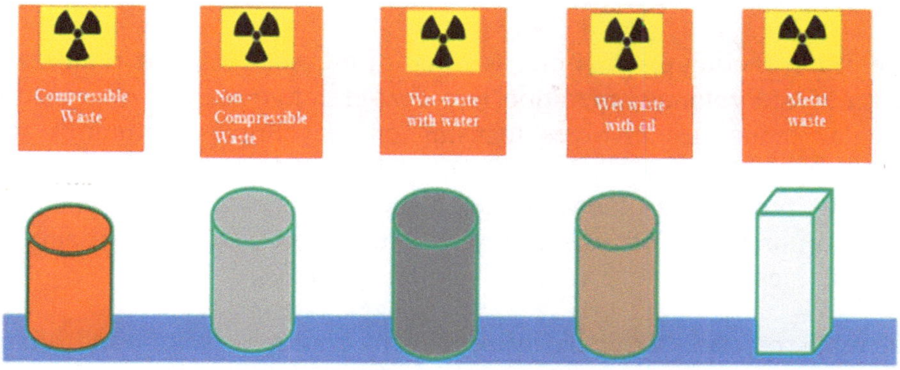

FIG. 32. Posters near the collection areas facilitating waste segregation. (Courtesy of Nuclear Safety Administration, Slovenia).

5.4. SELECTION OF PERSONAL PROTECTIVE EQUIPMENTS AND CONSUMABLES

A key component of a dry solid waste reduction programme is the selection of reusable, incinerable, and dissolvable products, such as personal protective equipment (PPE) and consumables. For PPE selection, the use of launderable PPE (Fig. 33) and washable protective clothing including coveralls, gloves, booties and launderable hoods (PAPRs), plays a significant role in the dry solid waste reduction programme.

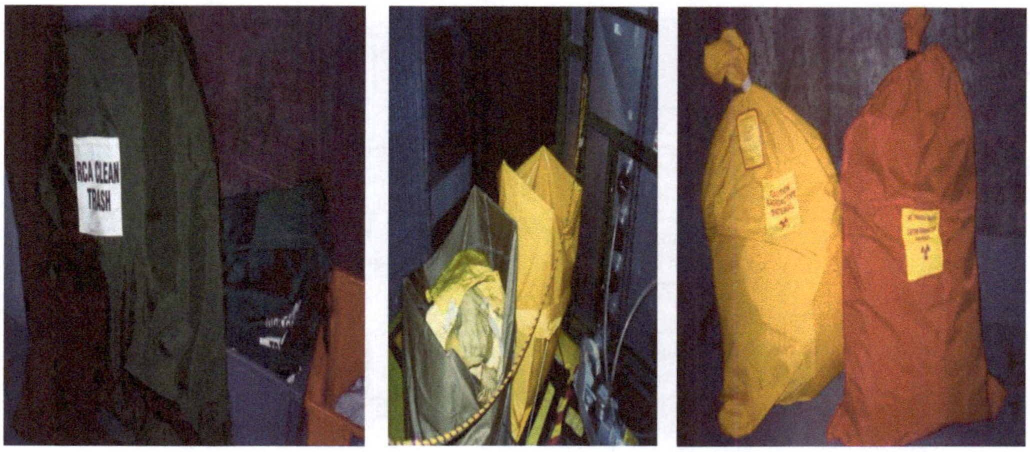

FIG. 33. Launderable bags for potentially clean and contaminated rubbish. (Courtesy of Pacific Gas & Electric Company, USA).

The use of launderable or easy to decontaminate consumable materials such as bags, tarps, sheeting for reusable insulation bags, tool and equipment bags, rubber area boundaries and scaffold storage bags, need to be taken into consideration. At Diablo Canyon NPP, reusable items have replaced materials that were previously classified as 'single use' only.

5.5. DISSOLVABLE MATERIALS

Temperature dissolvable poly-vinyl-alcohol (PVA) materials could substitute combustible materials such as clothing, rags, bags, and mops. PVA materials provide a high-volume reduction ratio through thermal destruction, using equipment that is less costly than incinerators or super compactors. However, the treatment cost of the resulting liquid product may influence the feasibility of implementation.

5.6. RECYCLE AND REUSABLE MATERIALS

The objective of a recycling programme, as a part of the waste volume minimization strategy, is to reduce the waste volume generation in a cost-effective manner. In waste management, fostering circular economies involves focusing on recycling opportunities such as proper segregation of recyclable items from waste, use of reusable materials, stringent controls for item replacement, processing of contaminated metal into cast iron containers or shielding components, incorporation of incineration with energy recovery, and utilizing modern waste recycling technologies to conserve valuable natural resources.

Recycle and reuse practices may be influenced by multiple factors and adopting a linear decision tree approach (Fig. 34) could facilitate the evaluation process for implementing a recycling programme.

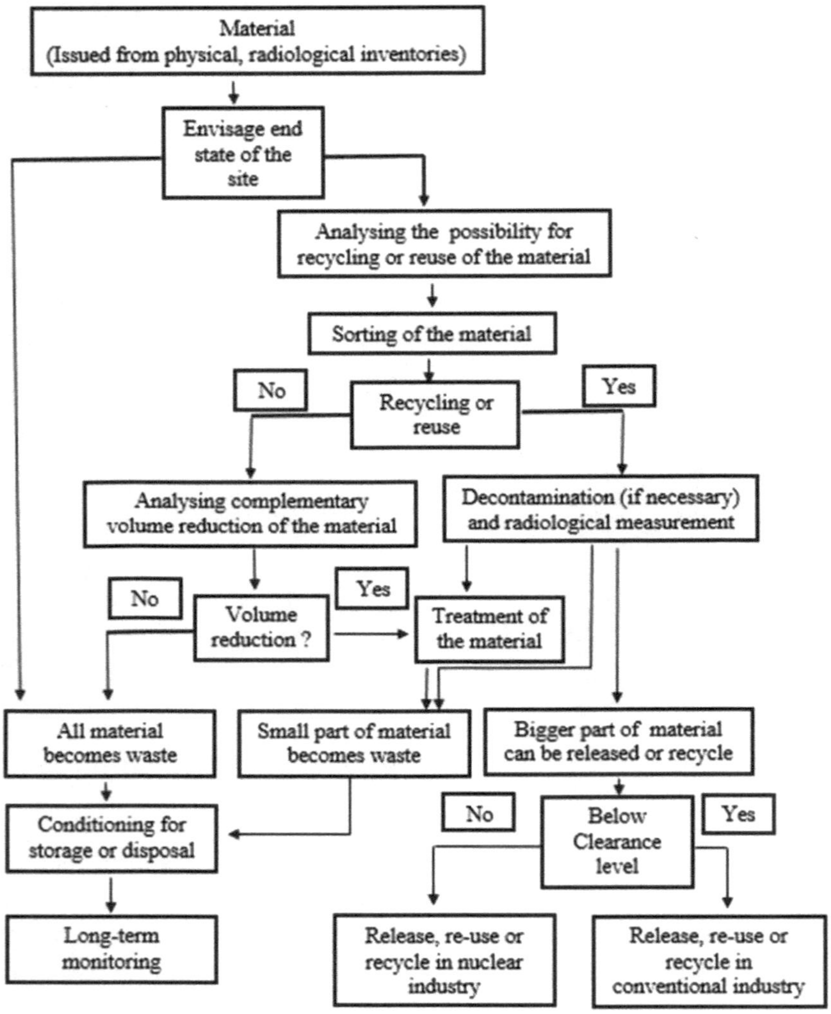

FIG. 34. Linear decision tree approach for waste recycling and reuse options. (Courtesy of EDF, France and World Nuclear Association (WNA)).

5.7. ADDITIONAL EXAMPLES FROM INDUSTRY

Other examples of dry solid waste reduction include the use of metal plates, foils and prefabricated metal modular walls to replace the use of plastic sheeting to cover floor grating and erect tents (Fig. 35). Metal plates and modular walls are reusable. Additionally, consider replacing porous wooden scaffolding with reusable metal scaffolding, which is easier to decontaminate.

FIG. 35. Reusable metal plate over floor grating. (Courtesy of Pacific Gas & Electric Company, USA).

Replacing poly-vinyl-chloride (PVC) plastic items with incinerable polyethylene (PE) products reduces procurement costs as PE is typically lower cost versus PVC (Fig. 36). Where visual clarity is required (e.g., bubble hoods, instrument bag covers) polyurethane (PU) offers an incinerable alternative to PVC. Similarly, rubber hoses can be replaced with braided PU tubing and PVC (commercially available Tygon tubing) can be replaced with incinerable PE tubing.

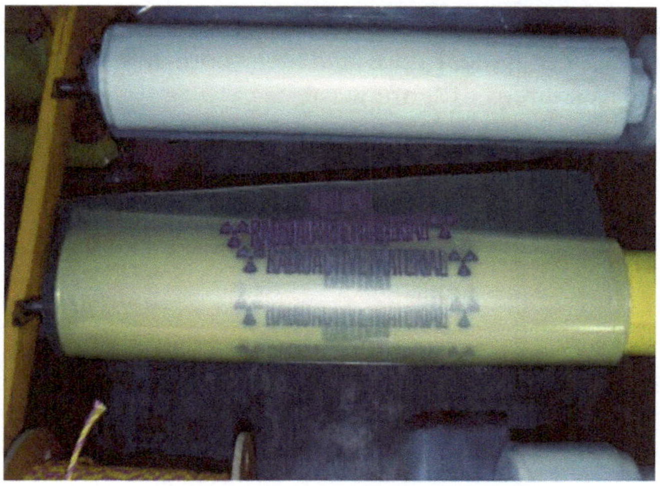

FIG. 36. Incinerable Polyethylene (PE) plastic bags. (Courtesy of Pacific Gas & Electric Company, USA).

6. INFLUENCE OF MATERIAL COMPOSITION ON WASTE GENERATION

6.1. OVERVIEW

To minimize activated waste, careful material selection during the design stage of reactor facilities is important. This selection affects both the economic and safety aspects of facility operation, as well as the volume of activated material requiring storage and disposal.

Radioactive waste, induced by the neutron activation of impurities within the reactor vessel, surrounding components and biological shielding is influenced by several factors:

- The amount of impurities in steels near the reactor vessel;
- The quantity of equipment and components in contact with the coolant;
- Additives introduced into the primary water system (water chemistry);
- Impurities in the fuel cladding material.

The level of activation can be minimized by reducing the impurities content. For example, ^{59}Co present in steels, core or other primary system components, can be activated to ^{60}Co, increasing the dose rate in areas surrounding those components.

6.2. NEUTRON ACTIVATION REACTIONS

Neutron-activated components represent the main contributors to radioactive inventories in core waste. Most of the core and near-core structures in light water reactors (LWRs) consist of ferrous alloys and concrete, while gas-cooled reactors use similar materials along with graphite. Therefore, when designing a NPP, the selection and use of materials with minimal concentrations of impurities susceptible to activation need to be carefully considered.

The reactor core is the part of a nuclear reactor that contains the nuclear fuel and is typically located inside a reactor vessel within the containment structure. The term ‘core waste’ refers to the radioactive waste generated during operation or midlife replacement (MLR) of the core, near-core structures and components which have been activated by neutron flux. Figure 37 [63] illustrates a cross-section of a typical core and near-core multiple regions highlighting the multiple areas that serve as sources of ‘core waste’.

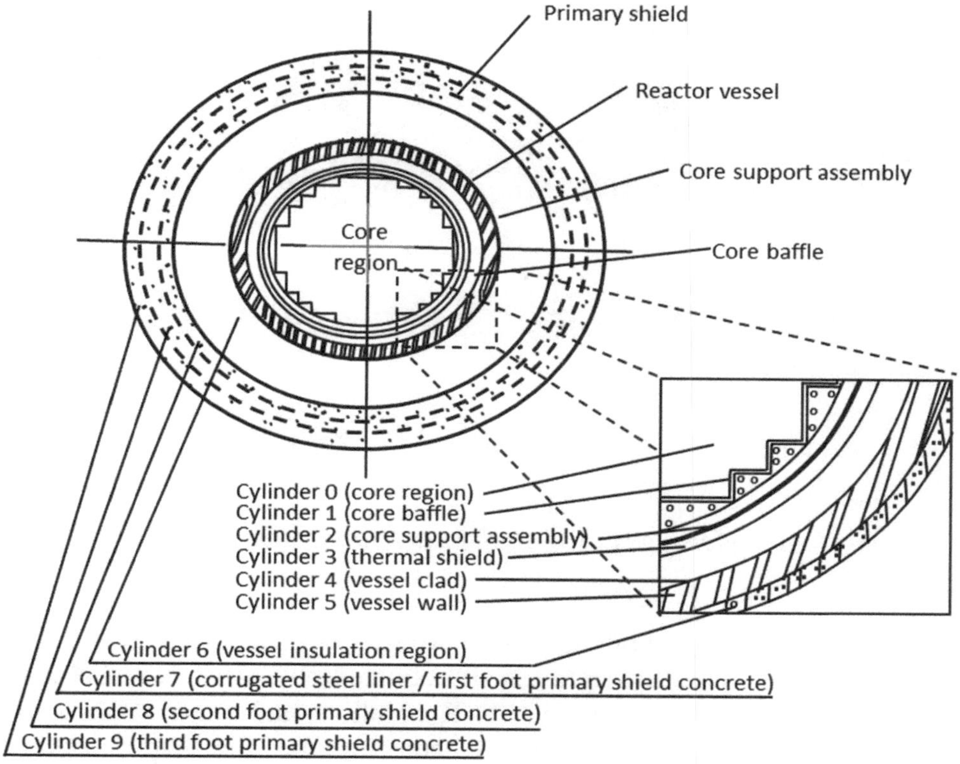

FIG. 37. Core and Near-Core Multiple Regions and Components (reproduced from [63]).

Examples of activation levels in major core and associated structural components are provided in Table I-2 in Appendix I for various reactor types, including boiling water reactors (BWR), pressurized water reactors (PWR), gas-cooled graphite-moderated reactors (GCR - the UK Magnox type and the French model), water-water energetic reactors (VVER), Canada Deuterium Uranium reactors (CANDU), pressurized heavy water reactors (PHWR), Training, Research, Isotope production and General Atomic reactor (TRIGA) and fast breeder reactors (FBR) [63].

6.3. MATERIAL SELECTION

6.3.1. Structural metals

The activation characteristics (based on thermal neutron flux of 1.0431 x 10^{15} $cm^{-2}s^{-1}$) can be classified in three categories: low, intermediate, and high probability, as shown in Fig. 38 [64].

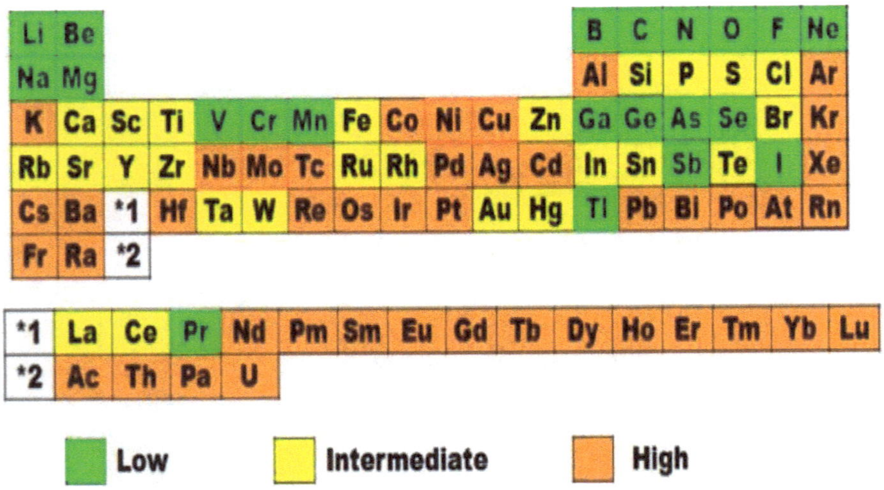

FIG. 38. Categorization of elements based on their activation characteristics (reproduced from [64]).

By reducing the accumulation of high activation probability elements such as Co, Nb, Cu, Cd, Mo, and Ag in core materials, lower levels of activated core waste can be achieved. Low activation structural materials, such as SiC, V-Ti-Cr alloys and martensitic steels (where W substitutes for Mo, and V and/or Ta for Nb) have been developed for this purpose. Additionally, ferritic steels and vanadium alloys are recognised as low activation structural alloys, while graphite and silicon carbide are considered low activation structural ceramics. The selection of these construction materials, currently used in PWR and LWR, sodium-cooled fast neutron and GFR/VHTR reactors is shown in Table 3 (adapted from [65]).

TABLE 3. CONSTRUCTION MATERIALS FOR CURRENT REACTOR DESIGNS [65]

	LWR/PWR	Sodium-cooled fast neutron reactor (FBR)	GFR/VHTR
Coolant	Water	Sodium	Helium
Temperature	288 – 360 °C	500 – 550 °C	550 – 1100 °C
Cladding	Zirconium-based	9 or 12Cr Steels	SiC
Core Internals	304/316 SS	316 SS	SiC or Alloy 800H
Vessel	Steel/316SS	316SS	Steel/316SS
Heat Exchanger	Alloy 600/690	9-12Cr/316 SS	Alloy 617
Piping	SS or LA Steel	9-12Cr/316 SS	Alloy 617

To reduce the activation of ^{60}Co from ^{59}Co, impurity limits for metals used in reactors are provided in Table 4.

TABLE 4. PROPOSED COBALT CONCENTRATION LIMITS IN REACTOR MATERIALS [66]

Material Content	Concentration limits (ppm)
Zircaloy-2	2-20
Zircaloy-4	2-20
Inconel-718	4694-10 000
Inconel X-750	4694-10 000
Carbon steels	300-10 000
Monel	260
Cupro-nickels	300
Stainless steel for high flux areas	500
Stainless steel 304	520

Due to advancements in metal development, cobalt-free substitute materials such as Pantanex 25, Actinit DUR 300, Nitronic 60, Tribaloy T 700, LC-1C, SS174 PH and Colmonoy 4 are now available. The effective use of these materials in nuclear environments is being investigated. An example from India is described in Annex III-2.

The Environment Agency in available techniques to prevent or minimize activated waste as proposed by the UK EPR, drawing on the operational experience of EDF and AREVA. These evaluations were part of their preconstruction environmental publications and supporting documents [66]. Key findings from the publication are as follows:

— ^{110m}Ag is found in control rods and helicoflex seals due to the activation of ^{109}Ag. The use of graphite seals could be considered as a replacement for helicoflex seals;
— ^{51}Cr is found in stainless steel due to the activation of ^{50}Cr. Since stainless steel is used for structural components in a reactor system and chromium is an essential element in stainless steel, prevention is not practical;
— ^{59}Fe and ^{54}Mn are found in stainless steel due to the activation of ^{58}Fe and ^{54}Fe respectively. As iron is an essential component of stainless steel, which is widely used for structural components in reactor systems, prevention is not practical; ^{63}Ni and ^{58}Co are activation products of ^{62}Ni and ^{58}Ni (^{58}Ni (n, p) $\rightarrow$^{58}Co) respectively in structural materials, particularly Alloy 690 used in the steam generator tubes and of the stainless steel used as reactor materials. Since these metallic materials are essential, prevention is not practical;
— ^{122}Sb, ^{124}Sb and ^{125}Sb are activation products of antimony isotopes found in alloys used for seals and pump bearings. To reduce this, EDF and AREVA have implemented the following measures:
  - Reducing the use of helicoflex seals by replacing them with graphite seals;
  - Using rotor stops and bearings that do not contain antimony.
— ^{60}Co, resulting from the activation of ^{59}Co (^{59}Co (n, γ)$\rightarrow$^{60}Co is present in steels and high cobalt alloys like stellite, which are used as valve seats. During operation, stellite-

coated surfaces release particles into the coolant, and the ^{59}Co content becomes activated in the reactor core. To minimize the cobalt levels, the following measures are suggested:

- Reducing the use of stellite parts in reactor coolant pumps;
- Specifying the use of low cobalt content stainless steels, welding materials and steam generator tubing.

6.3.2. Fuel cladding

The source of antimony in the reactor coolant and spent fuel pool liquid at Diablo Canyon NPP was determined to be from the activation of tin that present in the fuel cladding of Vantage V fuel from Westinghouse. Antimony is difficult to remove using standard anion resin and had an adverse impact on liquid effluent activity. As a result, expensive selective resins were procured to remove the antimony from spent fuel pool and radwaste system liquids.

When the plant transitioned to Zerlo fuel, the levels of antimony in the reactor coolant were reduced. Considering the use of fuel that does not generate antimony could further minimize the need for anion and selective resins required for its removal.

6.3.3. Reactor coolant pump seals

Initially, the Brazilian Angra 2 NPP had natural antimony (57.2 % of ^{121}Sb and 42.8 % of ^{123}Sb) incorporated in reactor coolant pump seals. Natural antimony could dissolve and transfer into the primary water circuit, where it has a high tendency to deposit on the fuel elements. This deposition leads to neutron activation, producing two radioactive isotopes, ^{122}Sb and ^{124}Sb (^{121}Sb (n, γ) ^{122}Sb and ^{123}Sb (n, γ) ^{124}Sb).

At plant shutdown, both absorbed activated and non-activated antimony isotopes were released into the primary water circuit, contributing to the radiation fields. The radiological characteristics of ^{122}Sb, particularly its short half-life (2.7 days), mitigate its impact. In contrast, ^{124}Sb, with a half-life of 60 days, may become a significant contributor to the dose rates and needs to be removed before maintenance work during outages.

The removal of antimony using anionic resins is a complex process that compromises resin's lifetime, necessitating more frequent replacements and consequently increasing the generation of radioactive solid waste. To minimize this waste arisings, reducing the amount of antimony as an impurity in reactor coolant pump seals could decrease the volume of spent anionic resin.

6.4. EVALUATION OF SELECTED MATERIALS

The appropriate design and quality of systems, structures, and components important to safety are vital for the construction and operation of NPPs. The IAEA publication [67] provides information on approaches to evaluating the suitability of commercial grade items for use in NPPs safety systems. The long-term operations of NPPs have greatly benefited from the use of commercial-grade products, and the near-term deployment of advanced reactor designs, including small modular reactors, is expected to largely leverage these products.

7. CURRENT MONITORING AND ASSESSING OF WASTE MINIMIZATION PROGRAMMES

7.1. OVERVIEW

The overarching concept for any minimization initiative is that all previously discussed elements (Section 2–6) need to be periodically assessed to develop a coordinated and balanced minimization programme. The duration of the analysis and the frequency of assessments must be tailored to the site's needs and regulatory requirements. Analysing the minimization programme to achieve predefined goals requires careful review and coordination of key factors that include technology readiness, regulatory compliance, environmental responsibility, occupational exposure, cost and secondary waste generation.

7.2. COST EVALUATION

The national policies and legislation, along with economic resources, will always establish boundaries for the waste management activities aimed at reducing waste generation.

Complicated economic conditions often lead to the reallocation of funding originally earmarked for waste minimization technologies. When sufficient storage or disposal capacity exists to accommodate waste, enhanced waste management programmes are often mistakenly viewed as a low priority.

The choice of waste minimization is usually justified through cost–benefit analyses that includes the following factors:

— The cost of retrieval and processing of stored legacy waste;
— Contingency funding for unforeseen events;
— The potential for reusing materials;
— Credits or benefits based on national policies promoting recycling and reuse.

In this context, authorities may promote important financial drivers of waste minimization to encourage options for recycling or reusing of materials. Financial incentives could be provided in relation to the decreasing amounts of radioactive waste generated, adhering to the accepted ‘the polluter pays’ principle.

7.3. STANDARDIZATION AND ASSESSMENT

Waste minimization plans are valuable tools for documenting, communicating, and tracking strategies to improve site performance. The following elements could be included in a waste minimization plan:

— Clearly defined goals (e.g., desired waste volumes);
— Near and long-term initiatives that will be undertaken to achieve the site’s goals, ensuring that each initiative meets specific, measurable, achievable, realistic, and timely (SMART) criteria;
— Owners, due dates, and management sponsors for each initiative;
— Cost estimates for completing each initiative. When possible, utilize site capital projects to fund waste minimization initiatives;
— Tracking of initiatives in a corrective action or work planning database, if available;

— Review the waste minimization plan periodically with senior management (annual reviews is a common review frequency);
— Benchmarking opportunities (e.g., industry conferences) included in the waste minimization plan.

The Institute for Nuclear Power Operations (INPO) developed performance indicators which established minimum levels of performance in key areas related to waste minimization. The aim was to enhance performance, reduce operating and maintenance costs and ensure continuous improvement. To measure the effectiveness of this initiative, INPO implemented data tracking and economic analyses, suggesting the following reportable indicators:

— Accessed contaminated area, expressed as a percentage of the total plant;
— Three-year rolling average waste generation volumes specified in cubic meters (m3).

Additionally, the Electric Power Research Institute (EPRI) developed the following:

— Computer programmes capable of data tracking for benchmarking comparisons;
— Guidelines for reporting:
  - o Dry solid waste;
  - o Liquid waste;
  - o Waste in storage.

Improvements in depicting any trend in waste generation can be observed by comparing a yearly plot (Fig. 39) and a 3-year average plot (Fig. 40) of the same waste stream.

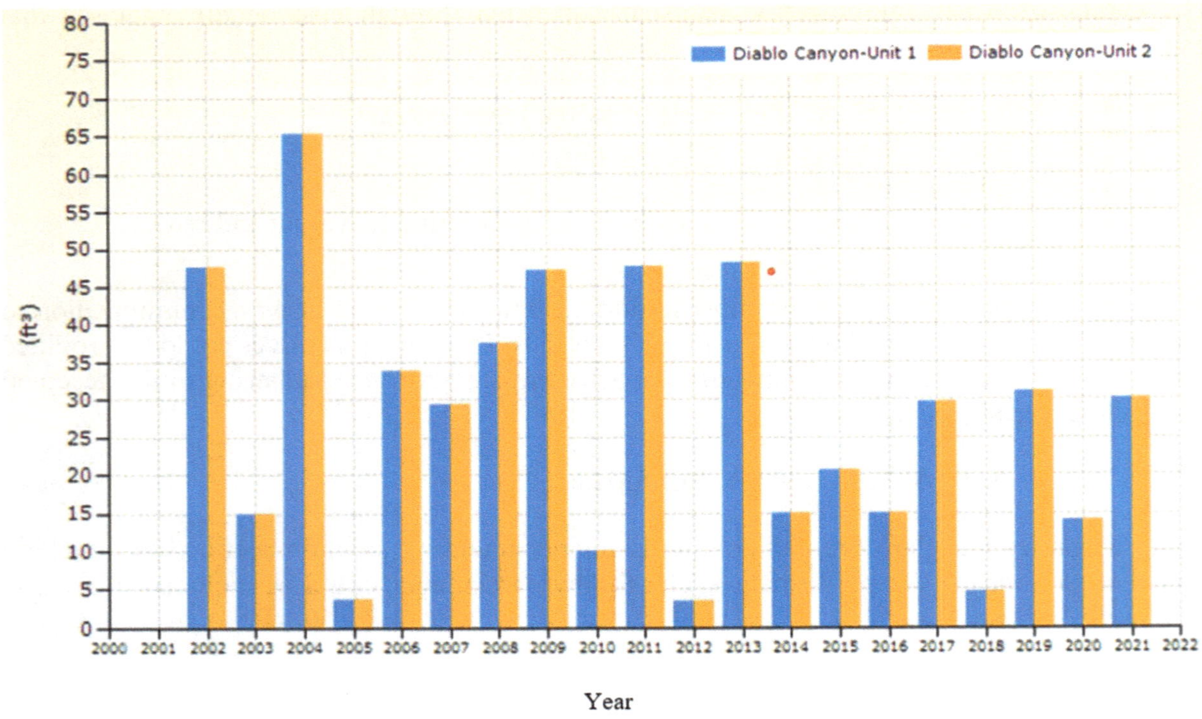

FIG. 39. Yearly average of secondary solid low and intermediate waste arising from treatment of liquid radioactive waste at Diablo Canyon NPP. (Courtesy of Pacific Gas & Electric Company, USA).

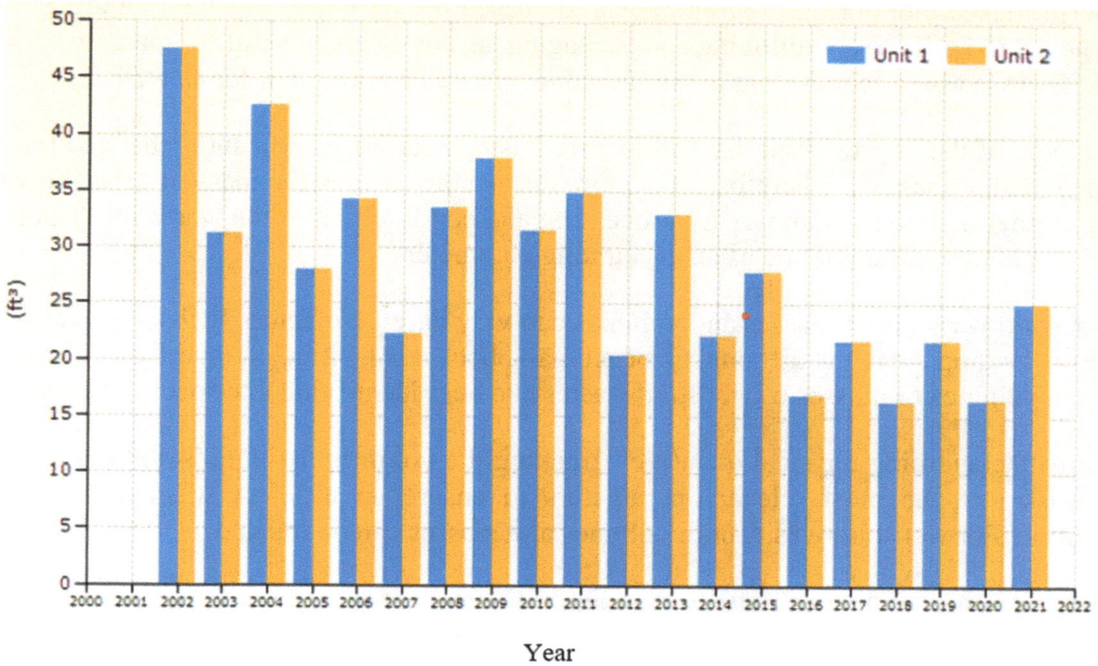

FIG. 40. Three-year average of secondary solid low and intermediate waste arising from treatment of liquid radioactive waste at Diablo Canyon. NPP. (Courtesy of Pacific Gas & Electric Company, USA).

7.4. MONITORING

The monitoring processes are designed to provide a logical path for evaluating an existing programme, management techniques and implementation considerations. They reflect a thorough analysis, optimization and coordination of the environmental, social, economic and technical aspects of a station's liquid, gaseous, and solid effluents streams and includes reactor-specific considerations. The process is divided into four distinct tasks:

— Evaluating the existing programme;
— Alternate management options;
— Implementation;
— Results and compliance verification.

Monitoring activities needs to be defined during the design and commissioning phases and are usually based on equipment functionality and media suppliers' recommendations. Typical parameters that could be monitored include:

— Pressure, flow, temperature;
— Chemical impurities, pH, conductivity;
— Activity and dose rate trends;
— Types and frequency of failures;
— Cost per unit of volume treated.

Performance monitoring could be an integral component of the management system's ongoing quality validation and safety case evaluations. Therefore, utilizing lead indicators instead of solely relying on reactive monitoring will help identifying safety and performance issues before they cause system problems.

The performance monitoring programme can be based on the volume of liquid and wet solid waste generated prior to conditioning and packaging, as this offers a useful comparison between plants. There is also a request that the liquid conductivity be accounted for as well.

The volume of the unprocessed dry solid waste arisings can be significantly changed through treatment and processing. Therefore, it is more reasonable to track the mass of dry solid waste arisings rather than the volume. After processing and conditioning of the waste for disposal, it is preferrable to use the volume as an accounting parameter.

For wet solid waste, both masses and volumes can vary based on factors such as water content and resin loading, making difficult to benchmark between facilities. Nevertheless, tracking either the volumes or masses of wet solid wastes can help identifying site specific trends.

In summary, regarding physical quantities, tracking the volumes for liquids and the mass for solid waste are the most relevant parameters to ensure an efficient waste minimization programme. For wet solid waste, both volumes and masses can be tracked.

The comparison of annual arisings over time, will reveal the trends. Due to the large quantity of waste from refuelling outages, a 3-year average provides a better comparison of performance as indicated in Fig. 41.

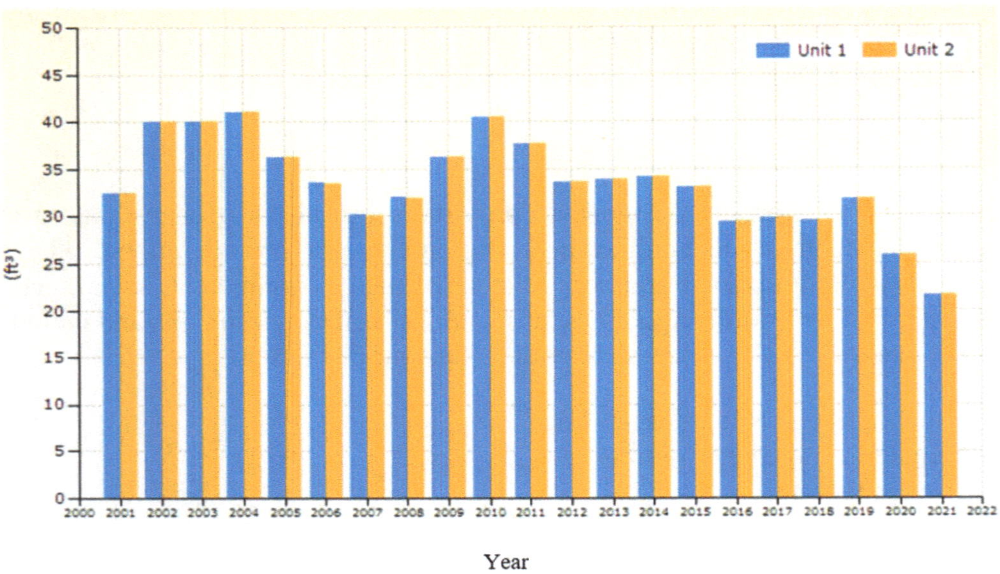

FIG. 41. Three-year average of Wet Solid Low and Intermediate Level Waste Arises at Diablo Canyon NPP. (Courtesy of Pacific Gas & Electric Company, USA).

7.5. BENCHMARKING

Benchmarking is a process which aims, in the same field of activity and purpose, to compare practices, products, methods, tools and expertise within different companies, entities, regulations and Member States, to facilitate the decision on what is most efficient or successful. Therefore, benchmarking has become a valuable means enabling a plant to improve its own performance while identifying potential gains in productivity, efficiency, profitability by assessing its own methodologies and tools and comparing it to current or innovative alternatives to be adapted or implemented instead.

While benchmarking may seem challenging for some utilities, especially when not all relevant data have been previously recorded, this should not deter them from undertaking the process. Even if all data are not immediately available at the outset, it is still possible to improve data collection incrementally. For example, some nuclear operators, such as the French EDF, have developed specific methodologies - such as "DOCADEC" - which incorporate on-site digitalization tools to determine, collect, and verify the necessary data, thereby enhancing the efficiency of benchmarking and waste management operations. In addition to evaluating the performance of a single plant, comparing average industry waste arisings at plants of similar design (as shown in Fig. 42) can also lead to improvements in waste management practices. Plants with lower waste arisings could be contacted to learn about their deployed practices and investigate whether they can be implemented or adapted at your plant.

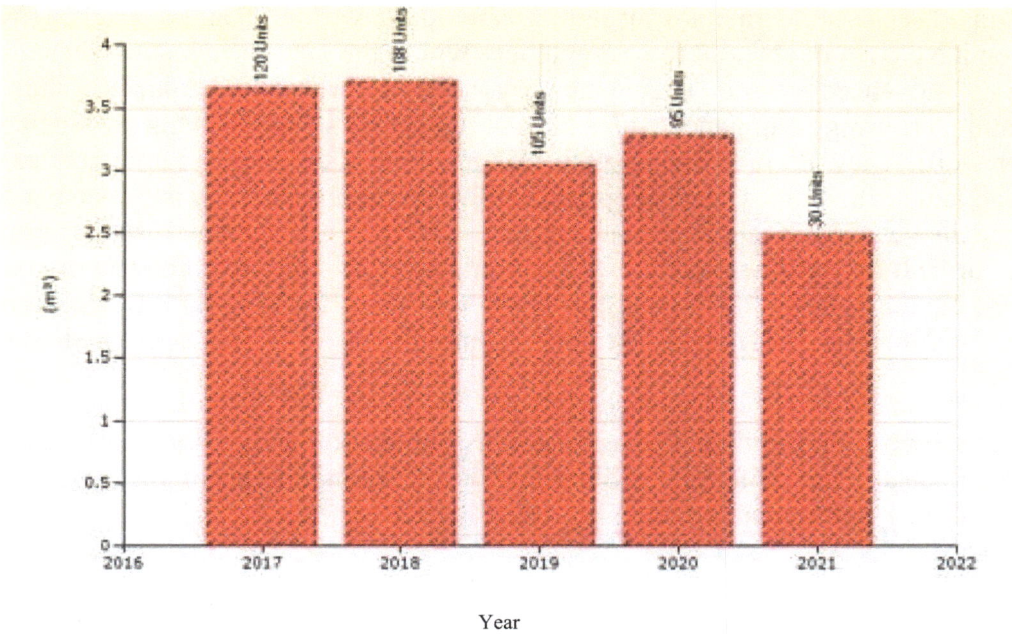

FIG. 42. Average industry wet solid intermediate level waste generated, 2017 – 2021 (m^3). (Source: EPRI RadBench™ Database).

However, to be representative, a sound benchmarking needs to follow an integrated approach. Effective benchmarking practices or processes not only focus on results (e.g., the efficiency of a waste cutting tool) but also consider various criteria and facets, such as cutting speed, the need of individual protective equipment, the volume of associated consumables and secondary waste, operators' dose exposure, easiness of use and settling the tool, environmental impacts, financial costs, requirements for prior operator qualification, training and duration of settlement. Consequently, identifying the most representative and suitable metrics for assessing practices, methodologies, and operators' expertise as well as their ability, is essential for the robustness of a benchmarking process.

In addition to representative criteria, it is crucial to establish a consistent basis for comparison, such as the conditions under which a particular methodology or process is applied. In the nuclear field, these conditions of applications are often dictated by local regulatory requirements, which can vary from one Member State to another.

Sharing experiences regarding improvements in R&D, along with implemented benchmarking processes and outcomes through international cooperation, could enhance the accuracy and relevance of benchmarking efforts.

8. CONCLUSIONS

While reactor sites are required to operate within a strict framework of legally binding regulations and license conditions, many operators go beyond compliance by actively seeking additional options to minimize and optimize the waste generated during the operation of Nuclear Power Plants (NPPs). This proactive approach has driven significant progress in the field of waste minimization, contributing to improved environmental performance and operational efficiency. These efforts extend to the design of new facilities, modifications to existing plants, and the implementation of effective decontamination strategies, all of which aim to reduce the volume and radioactivity of waste produced.

Demonstrated successful strategies for minimizing liquid, wet solid, and dry radioactive waste generation have become integral to best practices within the nuclear industry. These strategies incorporate advanced waste segregation, volume reduction techniques, and process optimization, ensuring compliance with stringent regulatory standards while enhancing overall operational efficiency. By prioritizing waste minimization at the source, implementing cutting-edge treatment technologies, and adopting a lifecycle approach to waste management, the industry achieves significant improvements in safety, environmental sustainability, and resource utilization. Furthermore, these initiatives contribute to broader organizational objectives, including cost reduction, regulatory compliance, and long-term waste management planning, reinforcing the nuclear sector's commitment to responsible and sustainable operations.

The waste minimization options discussed in this publication can be incorporated at various stages of a nuclear plant's lifecycle, from initial design through to decommissioning. Some of the key strategies highlighted include:

— Use of materials with low activation properties: Where feasible, selecting materials that are less prone to becoming radioactive when exposed to neutron flux reduces long-term waste generation. This consideration is especially crucial in the design phase, where the choice of structural materials can have a lasting impact on waste minimization.
— Avoiding contamination of the plant and equipment: Preventative measures to avoid contamination at its source can drastically reduce the need for extensive decontamination and waste handling later. These include the use of barriers and protective coatings.
— Ensuring proper separation between contaminated and non-contaminated areas: By maintaining strict separation, operators can prevent the unnecessary spread of contamination. This approach simplifies cleanup efforts and reduces the volume of waste designated as radioactive.
— Controlling radioactive materials within the plant: Effective control of radioactive materials, including careful handling, storage, and transfer, ensures that the spread of contamination is minimized, further limiting waste generation.
— Appropriate segregation of waste: By separating radioactive and non-radioactive materials, plants can limit the amount of material classified as radioactive, thereby reducing disposal costs and environmental impact.
— Minimizing the spread of contamination through containment measures: Employing effective containment strategies, such as sealed environments or the use of airlocks, helps confine radioactive materials to specific areas, reducing the need for extensive cleanup and disposal efforts.

— Reusing and recycling materials: Wherever possible, recycling and reusing materials can substantially reduce waste volumes. For example, decontaminated metal or equipment can often be reused within the plant, avoiding the need for disposal as radioactive waste.
— Preventing contamination by restricting material ingress into radioactive areas: By limiting the entry of non-essential materials into controlled or radioactive areas, plants can prevent unnecessary contamination, making decontamination and waste disposal more efficient.
— Assessing the application of technical advances in waste minimization: Ongoing assessment of emerging technologies and techniques can provide opportunities to further reduce waste volumes. Implementing these advances can enhance operational efficiency and contribute to long-term waste reduction goals.

This publication offers detailed technical information aimed at helping end-users apply these strategies effectively throughout the lifecycle of an NPP. It highlights the latest state-of-the-art practices and provides insights gained from reviewing industry experiences and evaluating real-world practices of waste minimization. By sharing these experiences, the publication aims to support the nuclear industry in achieving higher standards of waste management, contributing to safer, cleaner, and more sustainable operations.

APPENDIX I
ADDITIONAL INFORMATION ON CORE MATERIALS

I.1. NEUTRON ACTIVATION OF MATERIALS

Neutron-activated components represent, by far, the major contribution to waste radioactive inventories and the major neutron activation reactions and the activation products from elements (found in core materials) are provided in Table I.1 compiled from Ref [63].

TABLE I.1. IMPORTANT ACTIVATION REACTIONS FOR CORE AND NEAR-CORE MATERIALS [63].

Parent	Nuclear reaction (cross section)	Daughter nuclide	Principal emissions	Half-life of daughter (a)	Abundance of parent nuclide in parent element (%)	Comments
^{6}Li	n, α (953b)	^{3}H	β	12.3	7.5	Parent materials: concrete. Pure β emitter with maximum energy of 19.0 keV
^{13}C	n, γ (0.9mb)	^{14}C	β	5 730	1.1	Parent materials: primarily concrete and graphite Minor production route is with ^{13}C and ^{12}C
^{14}N	n, p (1.81b)	^{14}C	β	5 730	99.6	Parent materials: primarily concrete and graphite Pure β emitter with maximum energy of 156 keV
^{23}Na	n,2n	^{22}Na	$β^+$, EC	2.6	100	Parent materials: not significant for core waste
^{23}Na	γ, n	^{22}Na	$β^+$, EC	2.6	100	β emitter and γ with maximum energy of 1.275 MeV

TABLE I.1. IMPORTANT ACTIVATION REACTIONS FOR CORE AND NEAR-CORE MATERIALS [63] (CONT.).

Parent	Nuclear reaction (cross section)	Daughter nuclide	Principal emissions	Half-life of daughter (a)	Abundance of parent nuclide in parent element (%)	Comments
^{35}Cl	n, γ (2b)	^{36}Cl	β^- (β^+, EC)	301 000	75.8	Parent materials: graphite and in most steels as a trace element Principal decay is by β with maximum energy of 709 keV
^{40}Ca	n, γ	^{41}Ca	EC	103 000	96.9	Parent materials: concrete and impurity in graphite Principal decay is by EC with weak x-ray emission
^{54}Fe-54	n, p (53mb)	^{54}Mn	EC, γ	0.86	5.9	Parent materials: steel Principal decay is by EC with 835 keV γ emission
^{54}Fe	n, γ (2.25b)	^{55}Fe	EC, X	2.7	5.9	Parent materials: steel Principal decay is by EC with weak x-ray emission
^{58}Ni	n, γ (4.6b)	^{59}Ni	EC, X	76 000	68.3	Parent materials: austenitic stainless steel and nickel-based alloys Principal decay is by EC with continuous x-ray spectrum up to 1.07 MeV
^{62}Ni	n, γ (14.2b)	^{63}Ni	β^-	100	3.6	Parent materials: austenitic stainless steel and nickel-based alloys Principal decay is by β with maximum energy of 67 keV

TABLE I.1. IMPORTANT ACTIVATION REACTIONS FOR CORE AND NEAR-CORE MATERIALS [63] (CONT.).

Parent	Nuclear reaction (cross section)	Daughter nuclide	Principal emissions	Half-life of daughter (a)	Abundance of parent nuclide in parent element (%)	Comments
^{59}Co	n, γ (18.7b)	^{60}Co	β^-, γ	5.3	100	Parent materials: Co is a trace element in both carbon and stainless steels and in nickel-based alloys Principal decay is by β with maximum energy of 318 keV and two γ emissions of 1.17 MeV and 1.33 MeV
^{64}Zn	n, γ	^{65}Zn	EC, β^+	0.67	48.6	Parent materials: Zinc alloys, not typical core materials. Principle decay is by EC with β^+ with maximum energy of 329 keV
^{92}Zr	n, γ	^{93}Zr	β^-	1 500 000	17.1	Parent materials: Zirconium alloys Principal decay is by β with maximum energy of 60 keV. Decays to ^{93m}Nb, a γ emitter
^{92}Mo	n, γ	^{93}Mo	EC, X	3 500	14.8	Parent materials: stainless steels Principal decay is by EC to ^{93m}Nb and then to ^{93}Nb with a 19 keV γ.
^{93}Nb	n, γ	^{93m}Nb	IT, X	15.8	100	

TABLE I.1. IMPORTANT ACTIVATION REACTIONS FOR CORE AND NEAR-CORE MATERIALS [63] (CONT.).

Parent	Nuclear reaction (cross section)	Daughter nuclide	Principal emissions	Half-life of daughter (a)	Abundance of parent nuclide in parent element (%)	Comments
^{93}Nb	n, γ (1.15b)	^{94}Nb	β^-, γ	20 000	100	Parent materials: stainless steel Principal decay is by β with maximum energy of 472 keV to ^{94}Mo with a cascade of 871 keV and 703 keV and then to ^{93}Nb with a 19 keV γ. Additionally, ^{93m}Nb will be generated, which has a 16.1-year half-life and decays via internal transition, generating weak X rays. ^{94}Nb decays (half-life: 20 300 a) by b– emission (maximum energy: 472 keV) to a single level of ^{94}Mo at 1.574 MeV
^{93}Nb	n, γ	^{93m}Nb	IT, X	15.8	100	Parent materials: silver alloys used as control rods or Reactor Pressure Vessel (RPV) head gaskets Principal decay is by EC with weak X rays. ^{108}Ag decays by B with maximum energy of 1.655 MeV to stable ^{108}Cd the low cross-section (0.33 b) of ^{107}Ag and the long half-life of ^{108m}Ag limit its production

TABLE I.1. IMPORTANT ACTIVATION REACTIONS FOR CORE AND NEAR-CORE MATERIALS [63] (CONT.).

Parent	Nuclear reaction (cross section)	Daughter nuclide	Principal emissions	Half-life of daughter (a)	Abundance of parent nuclide in parent element (%)	Comments
^{94}Nb	n, p	^{94}Mo	β^-, γ	20 000	9.3	Parent materials: silver alloys used as control rods or RPV head gaskets Principal decay is by β with maximum energy of 1.467 MeV to ^{110}Ag with decays by β with maximum energy of 2.893 MeV to stable ^{110}Cd.
^{124}Sn	n, γ	^{125}Sb	β^-, γ	2.76	5.8	Principal decay is by β with maximum energy of 2.35 MeV to ^{125}Sb with decays by β with maximum energy of 622 keV to ^{125m}Te with 428, 600, 636, or 463 keV
^{132}Ba	n, γ	^{133}Ba	EC, X, γ	10.5	0.1	Parent materials: concrete Principal decay is by EC and γ emission to stable ^{133}Cs. Main γ energies are 356, 303, 383, and 276 keV.
^{133}Cs	n,γ	^{134}Cs	β , EC	2.065	100	Parent materials: concrete and fission product Principal decay is by β with maximum energy of 658 keV or by EC with and γ emission of 605 and 796 keV to stable ^{134}Ba.

TABLE I.1. IMPORTANT ACTIVATION REACTIONS FOR CORE AND NEAR-CORE MATERIALS [63] (CONT.).

Parent	Nuclear reaction (cross section)	Daughter nuclide	Principal emissions	Half-life of daughter (a)	Abundance of parent nuclide in parent element (%)	Comments
^{151}Eu	n,γ	^{152}Eu	EC, X, β^-, γ	13.5	47.8	Parent materials: concrete and core graphite
^{153}Eu	n,γ	^{154}Eu	β^-, X	8.6	52.2	Principal decay ^{152}Eu by β with maximum 1.477 MeV followed by α decay of ^{152}Gd
^{154}Eu	n,γ	^{155}Eu	β^-, , X	4.76	0	^{154}Eu decays by β with maximum energy of 1.85 MeV to stable ^{154} Gd. ^{155}Eu decays by β with maximum 2.52 keV to stable ^{155} Gd.
^{165}Ho	n,γ (66.5b)	^{166m}Ho	β^-, γ, X	1 200	100	Parent materials: impurity levels in steel and graphite. Principal decay is by β with maximum energy of 1.314 MeV followed by γ emission of 184, 810 and 712 keV.

I.2. INVENTORY OF CORE MATERIAL FROM DIFFERENT REACTORS

Examples of the radioactive inventory of reactor core, near-core components and associated structural components are presented in Table I-2 (compiled from Ref [64]) and include information for boiling water reactor (BWR), pressurized water reactor (PWR), gas-cooled graphite-moderated reactor (GCR-the UK Magnox type and the French model), water-water energetic reactor (VVER), Canada deuterium uranium reactor (CANDU), pressurized heavy water reactor (PHWR), training, research, isotope production and general atomic reactor (TRIGA) and fast breeder reactors (FBR).

TABLE I.2. RADIOACTIVE INVENTORY OF REACTOR CORE AND NEAR-CORE COMPONENTS [64].

Reactor Type and Conditions	Components	Calculated Radioactivity [Bq]
BWR 2590 MW (th), 7 years of irradiation, 4 effective full power years (EFPY), 5 years after shutdown (Caorso, Italy)	Internals	1.28E+16
	Fuel cases	2.44E+15
	Control rods	3.40E+15
	Reactor pressure vessel	3.84E+13
	Sacrificial shield	6.18E+11
	Biological shield	3.33E+09
	Dry well	1.51E+10
PWR 870 MW (th), 23 years of irradiation, 10.6 EFPY, 5 years after shutdown. (Trino, Italy)	Internals	4.27E+15
	Control rods	1.16E+15
	Vessel	3.52E+14
	Neutron shield	2.47E+12
	Biological shield	9.39E+09
GCR (Magnox) 355 MW (th), 26 years of irradiation, 5 years after shutdown. (UK)	Reactor pressure vessel (metal)	8.30E+15
	Reactor internal structures (metal)	4.50E+16
	Moderator (graphite)	4.40E+14
	Reflector (graphite)	1.50E+14
	Biological shield (concrete and steel reinforcing)	2.20E+13

TABLE I.2. RADIOACTIVE INVENTORY OF REACTOR CORE AND NEAR-CORE COMPONENTS [64] (CONT.)

Reactor Type and Conditions	Components	Calculated Radioactivity [Bq]
GCR (Électricité de France (EDF) 1830 MW (th), 21 years of irradiation, 14 EFPY, 6 years after shutdown. (St. Laurent, France)	Graphite (moderator + reflector)	2.70E+15
	Internal structures (steel)	1.10E+16
	Casing	8.90E+13
	Biological shield	6.00E+14
VVER 440 1350 MW (th), 17 years of irradiation, 9 EFPY (approx.), 6 years after shutdown. (Greifswald Unit 1, Germany)	Reactor pressure vessel	1.00E+14
	Protection tube unit	9.40E+14
	Core basket	1.00E+16
	Reactor cavity with cavity bottom	2.00E+15
	Annular water tank	7.00E+12

TABLE I.2. RADIOACTIVE INVENTORY OF REACTOR CORE AND NEAR-CORE COMPONENTS [64] (CONT.).

Reactor Type and Conditions	Components	Radioactivity [Bq]
CANDU NPD 98.5 MW (th), 21 years of irradiation, 17.1 EFPY, 5 years after shutdown. (Canada)	Calandria tube	5.57E+13
	Coolant tube	3.23E+14
	Inner Calandria shell	5.70E+13
	Inner Calandria end walls	1.06E+13
	Outer Calandria end wall	1.07E+10
	End fitting	1.60E+14
	Outer calandria shell	1.47E+09
	Radial concrete	5.68E+10
	End reflector stepped tube	5.66E+12
	Tube end support	2.97E+12
	Fuel latch	9.41E+11
	Closure plug assembly	3.55E+10
	Spacer sleeve	1.39E+13
KKN Heavy water-moderated, CO_2-cooled 320 MW (th), 2 years of irradiation, 10 EFPD (approx.), 5 years after shutdown. (Niederaichbach, Germany)	Reactor internals	1.20E+14
	Moderator tank	1.80E+13
	Thermal shield	8.00E+11
	Biological shield	1.30E+10

TABLE I.2. RADIOACTIVE INVENTORY OF REACTOR CORE AND NEAR-CORE COMPONENTS [64] (CONT.).

Reactor Type and Conditions	Components	Radioactivity [Bq]
PHWR 200 MW (th), 18 years of irradiation, 10 EFPY (approx.), 5 years after shutdown. (MZFR, Germany)	Reactor pressure vessel lid	4.20E+07
	Pressure vessel (without internals)	9.40E+12
	Moderator tank	2.80E+16
	Lagging	8.00E+11
	Lower spacer	4.10E+14
	Thermal shield	6.60E+14
	Fuel element channels	1.50E+14
	Upper spacer	6.80E+14
	Control rod guide tubes	6.50E+13
	Spacer ring	6.50E+13
TRIGA TRIGA Mark-II reactor 42 years at nominal power of 250 kW (Vienna, Austria)	Activated materials Steel Aluminium Concrete Graphite	 2.30E+13 6.30E+11 8.20E+10 3.40E+10
	Contaminated materials Steel Aluminium Small components with low activity	 2.70E+08 9.40E+07 3.50E+07

TABLE I.2. RADIOACTIVE INVENTORY OF REACTOR CORE AND NEAR-CORE COMPONENTS [64] (CONT.).

Reactor Type and Conditions	Components	Radioactivity [Bq]
FBR (Rapsodie) After 10 years of shutdown at nominal power of 20 MW (Cadarache, France)	Neutron shields	6.70E+12
	Thermal shields	0.30E+12
	Vessels	0.40E+12
	Diagrids	0.30E+12
	Core plate cover	0.70E+12
	Stellite-coated plates and brushings	28.4E+12
	Rare earth doped concrete	40.1E+12

REFERENCES

[1] INTERNATIONAL ATOMIC ENERGY AGENCY, IAEA Nuclear Safety and Security Glossary, Non-serial Publications, IAEA, Vienna (2022).

[2] JOINT CONVENTION ON THE SAFETY OF SPENT FUEL MANAGEMENT AND ON THE SAFETY OF RADIOACTIVE WASTE MANAGEMENT, INFCIRC/546 (1997), https://www.iaea.org/sites/default/files/infcirc546.pdf

[3] INTERNATIONAL ATOMIC ENERGY AGENCY, Safety of Nuclear Power Plants: Design, IAEA Safety Standards Series No. SSR-2/1 (Rev. 1), IAEA, Vienna (2016).

[4] INTERNATIONAL ATOMIC ENERGY AGENCY, Classification of Radioactive Waste, IAEA Safety Standards Series No. GSG-1, IAEA, Vienna (2009)

[5] HAJDÚ, D., DIAN, E., GMÉLING, K., KLINKBY, E., COOPER-JENSEN, C.P., OSÁN, J., ZAGYVAI, P., Experimental study of concrete activation compared to MCNP simulations for safety of neutron sources, Applied Radiation and Isotopes, 171 (2021) 1-10,https://doi.org/10.1016/j.apradiso.2021.109644

[6] EUROPEAN COMMISSION, FOOD AND AGRICULTURE ORGANIZATION OF THE UNITED NATIONS, INTERNATIONAL ATOMIC ENERGY AGENCY, INTERNATIONAL LABOUR ORGANIZATION, OECD NUCLEAR ENERGY AGENCY, PAN AMERICAN HEALTH ORGANIZATION, UNITED NATIONS ENVIRONMENT PROGRAMME, WORLD HEALTH ORGANIZATION, Radiation Protection and Safety of Radiation Sources: International Basic Safety Standards, IAEA Safety Standards Series No. GSR Part 3, IAEA, Vienna (2014).

[7] INTERNATIONAL ATOMIC ENERGY AGENCY, Regulatory Control of Radioactive Discharges to the Environment, IAEA Safety Standards Series No. GSG-9, IAEA, Vienna (2018)

[8] UNITED STATES DEPARTMENT OF ENERGY (DOE), "Annual Report of Waste Generation and Pollution Prevention Progress 1998", Report DOE/EM-0464 (1999).

[9] OSPAR COMMISSION, Convention for the Protection of the Marine Environment in the North-East Atlantic, OSPAR (1992).

[10] HELSINKI COMMISSION BALTIC MARINE ENVIRONMENT PROTECTION COMMISSION, The Convention on the Protection of the Marine Environment of the Baltic Sea, HELCOM, Helsinki (1992).

[11] UNITED NATIONS ENVIRONMENT PROGRAMME, Convention for the Protection of the Marine Environment and the Coastal Region of the Mediterranean, UNEP, Barcelona (1976).

[12] UNITED NATIONS ECONOMIC COMMISSION FOR EUROPE, Convention on Environmental Impact Assessment in a Transboundary Context (UNECE), Espoo (1991).

[13] UNITED NATIONS ECONOMIC COMMISSION FOR EUROPE, The Protocol on Strategic Environmental Assessment - Kyiv Proto col, Kyiv (2003).

[14] INTERNATIONAL ATOMIC ENERGY AGENCY, Radioactive Waste Management, Non-serial Publications, IAEA, Vienna (2001).

[15] NATIONAL NUCLEAR SAFETY ADMINISTRATION (NNSA), Waste Minimization for Nuclear Facilities, HAD 401–0, USA (2016).

[16] HUICHUN, L., China National Nuclear Power Co., Waste goals for outage Dry Active Waste (DAW), Private communication (2022).

[17] WORLD ASSOCIATION OF NUCLEAR OPERATORS (WANO), Performance Indicators 2012, 2204-WANO-Performance-Indicator-Electronic-Document-6pp_English.pdf.aspx

[18] MILLER, C., Pacific Gas & Electric Company (USA) Private communication (2024).

[19] INTERNATIONAL ATOMIC ENERGY AGENCY, Factors Affecting Public and Political Acceptance for the Implementation of Geological Disposal, IAEA-TECDOC-1566, IAEA, Vienna (2007).

[20] INTERNATIONAL ATOMIC ENERGY AGENCY, Considerations for Waste Minimization at the Design Stage of Nuclear Facilities, Technical Reports Series No. 460, IAEA, Vienna (2007).

[21] INTERNATIONAL ATOMIC ENERGY AGENCY, Leadership, Management and Culture for Safety in Radioactive Waste Management, IAEA Safety Standards Series No. GSG-16, IAEA, Vienna (2022)

[22] INTERNATIONAL ATOMIC ENERGY AGENCY, Safety Assessment for Facilities and Activities, IAEA Safety Standards Series No. GSR Part 4 (Rev. 1), IAEA, Vienna (2016).

[23] INTERNATIONAL ATOMIC ENERGY AGENCY, Predisposal Management of Radioactive Waste from Nuclear Power Plants and Research Reactors, IAEA Safety Standards Series No. SSG-40, IAEA, Vienna (2016).

[24] INTERNATIONAL ATOMIC ENERGY AGENCY, Predisposal Management of Radioactive Waste from Nuclear Fuel Cycle Facilities, IAEA Safety Standards Series No. SSG-41, IAEA, Vienna (2016).

[25] INTERNATIONAL ATOMIC ENERGY AGENCY, Storage of Radioactive Waste, IAEA Safety Standards Series No. WS-G-6.1, IAEA, Vienna (2006).

[26] INTERNATIONAL ATOMIC ENERGY AGENCY, Application of the Concepts of Exclusion, Exemption and Clearance, IAEA Safety Standards Series No. RS-G-1.7, IAEA, Vienna (2004).

[27] INTERNATIONAL ATOMIC ENERGY AGENCY, Safety of Nuclear Power Plants: Design, IAEA Safety Standards Series No. SSR-2/1 (Rev 1), IAEA Vienna (2016).

[28] INTERNATIONAL ATOMIC ENERGY AGENCY, Operational Limits and Conditions and Operating Procedures for Nuclear Power Plants, IAEA Safety Standards Series No. NS-G-2.2, IAEA, Vienna (2000).

[29] INTERNATIONAL ATOMIC ENERGY AGENCY, Radiation Protection Aspects in the Design of Nuclear Power Plants, IAEA Safety Standards Series No. NS-G-1.13, IAEA, Vienna (2005).

[30] INTERNATIONAL ATOMIC ENERGY AGENCY, Safety of Nuclear Power Plants: Commissioning and Operation, IAEA Safety Standards Series No. SSR-2/2 (Rev. 1), IAEA, Vienna (2016).

[31] INTERNATIONAL ATOMIC ENERGY AGENCY, Occupational Radiation Protection, IAEA General safety guide No. GSG-7, IAEA, Vienna (2018).

[32] INTERNATIONAL ATOMIC ENERGY AGENCY, Design of the Reactor Core for Nuclear Power Plants, IAEA Safety Standards Series No. SSG-52, IAEA, Vienna (2019).

[33] ROSS, M., Exelon Generation Radwaste Oversite and Management, Low-Level Waste Advisory Committee Meeting (2017), 6 Exelon Presentation rev1.pptx (live.com)

[34] TOKYO ELECTRIC POWER COMPANY HOLDINGS (TEPCO), Introduction of the dose and radioactive waste reduction program in Kashiwazaki-Kariwa NPS, atc2018_5-3.pdf (nsra.or.jp)

[35] ÉLECTRICITÉ DE FRANCE (EDF), Reaction of a new "DP2D",
ELECTRICITE DE FRANCE, EDF (inno4graph.eu)

[36] SOGIN GROUP, Protect the present guarantee the future, (2014), https://www.sogin.it/en

[37] INNO4GRAPH, ENRESA – Empresa Nacional de Residuos Radiactivos, https://www.inno4graph.eu/partners/enresa-empresa-nacional-de-residuos-radiactivos

[38] INTERNATIONAL ATOMIC ENERGY AGENCY, Minimization and segregation of radioactive wastes, IAEA-TECDOC-652, IAEA, Vienna (1992).

[39] INTERNATIONAL ATOMIC ENERGY AGENCY, Development and Implementation of a Process Based Management System, IAEA Nuclear Energy Series No. NG-T-1.3, IAEA, Vienna (2015)

[40] INTERNATIONAL ATOMIC ENERGY AGENCY, Leadership and Management for Safety, IAEA Safety Standards Series No. GSR Part 2, IAEA, Vienna (2016), https://doi.org/10.61092/iaea.cq1k-j5z3

[41] INTERNATIONAL ATOMIC ENERGY AGENCY, Invitation and Evaluation of Bids for Nuclear Power Plants, IAEA Nuclear Energy Series No. NG-T-3.9, IAEA, Vienna (2011).

[42] INTERNATIONAL ATOMIC ENERGY AGENCY, Guidance for Optimizing Nuclear Power Plant Maintenance Programmes, IAEA-TECDOC-1383, IAEA, Vienna (2003).

[43] NUCLEAR ENERGY AGENCY, Status, Barriers and Cost-Benefits of Robotic and Remote Systems Applications in Nuclear Decommissioning and Radioactive Waste Management, Report NEA/RWM/R (2022)1 (2023).

[44] FARKAS, P., Slovenian Nuclear Safety Administration, Private communication (2024).

[45] NUCLEAR ENGINEERING INTERNATIONAL, How gel formulations can aid decontamination, (2000), How gel formulations can aid decontamination - Nuclear Engineering International (neimagazine.com)

[46] INTERNATIONAL ATOMIC ENERGY AGENCY, Decontamination Methodologies and Approaches, IAEA Nuclear Energy Series No. NW-T-1.38, IAEA, Vienna (2023).

[47] CBI POLYMERS, Decongel Overview (2012), Decongel Overview | PPT (slideshare.net)

[48] NUCLEAR STREET NEWS TEAM, TEPCO Spraying Decontamination Gel at Fukushima Daiichi Nuclear Power Station – Video (2011), TEPCO Spraying Decontamination Gel at Fukushima Daiichi Nuclear Power Station – Video - News - Nuclear Power News - Nuclear Street - Nuclear Power Plant News, Jobs, and Careers

[49] YANG, H., YOON, I., LEE, Y., Poly (vinyl alcohol)-borax complex-based spray coating for the decontamination of radioactive Cs from wide-area surfaces. Chem. Eng. J., 402 (2020), https://doi.org/10.1016/j.cej.2020.126299

[50] MOORE, J.J., RAINE, T.P., JENKINS, A., LIVENS, F.R., LAW, K.A., MORRIS, K., LAW, G., YEATES, S.G., Decontamination of caesium and strontium from stainless steel surfaces using hydrogels, React. Funct. Polym., 142 (2019) 7-14, https://doi.org/10.1016/j.reactfunctpolym.2019.04.004

[51] YANG, H., PARK, C.W., LEE, K., Enhanced surface decontamination of radioactive Cs by self-generated, strippable hydrogels based on reversible cross-linking, J. Hazard. Mater. 362 (2019) 72-81, https://doi.org/10.1016/j.jhazmat.2018.08.064

[52] WANG, J., LIU, G., ZHAO, L., WANG, S., LI, J., Research on nuclear emergency decontamination technology based on strippable coating, J. Radioanal. Nucl. Chem. 322 (2019) 1049-1054, https://doi.org/10.1007/s10967-019-06790-5

[53] POZO, A.A.P., MONROY-GUZMÁN, F., GÓMORA-HERRERA, D.R., NAVARRETE-BOLAÑOS, J., BUSTOS, E.B., Radioactive decontamination of metal surfaces using peelable films made from chitosan gels and chitosan/magnetite nanoparticle composites, Prog. Nucl. Energy 144 (2022) 104088, DOI:10.1016/j.pnucene.2021.104088

[54] Nuclear News Wire, SuperGel to the Rescue (2022), SuperGel to the Rescue -- ANS / Nuclear Newswire

[55] GOSSARD, A., LILIN, A., FAURE, S., Gels, coatings and foams for radioactive surface decontamination: State of the art and challenges for the nuclear industry, Progress in Nuclear Energy 149 (2022), https://doi.org/10.1016/j.pnucene.2022.104255.

[56] GOSSARD, A., FRANCES, F., VENDITTI, P., LEPEYTRE, C., GRANDJEAN, A., The vacuumable gel technology for the nuclear decontamination of solid surfaces state of the art and current developments for decommissioning operations, GLOBAL 2017 International Nuclear Fuel Cycle Conference, Seoul, South Korea (2017).

[57] FOURNEL, B., FAURE, S., POUVREAU, J., DAME, C., POULAIN, S., Decontamination using foams: A brief review of 10 years French experience. In Proceedings of the 9th International Conference on Radioactive Waste Management and Environmental Remediation, Oxford, UK (2003), https://doi.org/10.1115/ICEM2003-4526

[58] INTERNATIONAL ATOMIC ENERGY AGENCY, New methods and techniques for decontamination in maintenance or decommissioning operations. Results of a coordinated research programme 1994-1998, TECDOC-1022, IAEA, Vienna (1999).

[59] INTERNATIONAL ATOMIC ENERGY AGENCY, Processing of Nuclear Power Plant Waste Streams Containing Boric Acid, IAEA-TECDOC-911, IAEA, Vienna (1996).

[60] KHAN, A.W., Chashma Nuclear Power (Pakistan), Private contribution as co-author (2024).

[61] ENERGYSOLTIONS, Bear Creek Processing Facility, Oak Ridge, Bear Creek Processing Facility - Energy Solutions

[62] BERRY. J.B., The Oak Ridge National Laboratory green is clean -Achieving waste reduction and efficiency goals, WM99 Conference (1999), https://citeseerx.ist.psu.edu/document?repid=rep1&type=pdf&doi=25b2eb4727a5a79314cab07fdd267458afb0f113

[63] INTERNATIONAL ATOMIC ENERGY AGENCY, Radiological Characterization of Shut Down Nuclear Reactors for Decommissioning Purposes, IAEA Technical Reports Series No. 389, IAEA, Vienna (1998).

[64] INTERNATIONAL ATOMIC ENERGY AGENCY, Handbook on Nuclear Activation Data, Technical Reports Series No. 273, IAEA, Vienna (1987).

[65] BUSHBY, J.T., Challenges for Reactor Materials, Nano Nuclear Workshop, Rice University (2012), https://www.tms.org/meetings/2012/nanonuclear/pdfs/background/BusbyReactorMaterialsChallenges.pdf

[66] GREEN, R., UK EPRTM nuclear power plant design by AREVA NP SAS and Electricité de France SA, GDA Final Assessment Report UK EPR-03 (2011).

[67] INTERNATIONAL ATOMIC ENERGY AGENCY, Suitability Evaluation of Commercial Grade Products for Use in Nuclear Power Plant Safety Systems, IAEA-TECDOC-2034, IAEA, Vienna (2023).

ANNEX I
INTERNATIONAL STUDIES ON LIQUID WASTE MINIMIZATION

The examples provided in the following pages address the organization and detailed technical aspects of examples of international case studies on liquid waste minimization at NPPs. The descriptions provide practical information on projects are planned and managed in various Member States. The examples given in this Section are not necessarily best practices but rather reflect a variety of international practices. These case studies reflect the experience and views of their contributors and, although consistent with the main text, are not for guidance.

I–1. SLOVENIA CASE STUDY

The Krško NPP, a joint project of Slovenia and Croatia, is a Westinghouse two-loop pressurized light water reactor (PWR) with nominal output power of 727/696 Mwe. It is in commercial operation from 1983 and could operate to 2043 pending the successful conclusion of Periodic Safety Reviews in 2023 and 2033.

The decommissioning of a NPP requires facilities for handling radioactively contaminated material - balancing the flow of material before its subsequent processing. As part of the decommissioning process of the V1 NPP, based on a feasibility study, the Nuclear and Decommissioning Company, Inc. (JAVYS, a.s.) was contracted to facilitate to decommissioning sections at the Krško NPP. JAVIS chose for this purpose the premises in which the collection and storage tanks for liquid and wet solid RAW were placed. They were located in the auxiliary building CB 801A:V1, but it was necessary to slightly modify them.

The modification considered removing these tanks. That is, gradually cleaning them (Fig I–1), prior to dismantling decontamination, segmentation into pieces of defined dimensions that can be handled in standard pallets and conform with standard already set RAW management processes.

FIG. I–1. Removing of solids (cleaning) the inside of spent liquid storage tanks. (Courtesy of JAVIS, Slovakia).

In the process of planning these activities, after the cleaning of the respective tanks, in view of their expected and later confirmed condition, a method of decontamination of the internal surfaces of the tanks by applying DeconGel was selected through a feasibility study, which significantly reduced the formation of secondary liquid waste.

Decontamination activities began by cutting a utility access hole in the side of the tank, allowing workers access to the inside of the tank. The cut off part of the tank hole as well as the bottom of the tank were sprayed with DeconGel (Fig. I–2 (a)) using an airless high-pressure machine. The device was placed outside the tank to avoid its contamination. After drying of the gel (within 24 hours), in which the contamination was fixed into the polymer matrix, the gel was manually pulled away from the cut entry utility access hole.

At the bottom of the tank, the dried gel was left as a layer preventing the possible spread of contamination and was pulled down as the last step of the decontamination work. The walls of the tank were covered with a layer of decontamination gel. The dried gel was removed (Fig. I–2 (b)) from the treated surfaces from top to bottom, and at the initial point of removal, the gel layer was cut and lifted with a knife. It was handled from a non-contact side, while workers were equipped with prescribed PPE (protective equipment). The torn gel was folded and stored directly in plastic bags (Tatrafol) and then in drums that were placed at the workplace.

Based on dry abrasions taken from the treated surfaces, which demonstrated the level of decontamination, in the case of a fixed contamination level higher than 0.3 Bq/cm^2, a second layer of gel was applied to the surfaces. After tearing off the second layer of gel, the dry abrasions were taken again.

a. Application of DeconGel *b. Removal of dried DeconGel*

FIG. I–2. Application of DeconGel and removing of formed gel layer the inside spent liquid storage tanks. (Courtesy of JAVIS, Slovakia).

In the event of residual contamination or hot spots, these were manually wiped with decontamination wipes impregnated with Alcatum, which complies with products and materials used in NPPs that are free of fluorides, chlorides and other compounds that could damage the primary circuit). A reduction in the level of fixed contamination below 0.3 Bq/cm^2 was demonstrated by abrasions taken.

The welds (including their immediate surroundings) that connect the stainless-steel plates forming the body of the tank were identified as very problematic places for decontamination. It is natural that the weld joint and transition layer to the base material may contain some level of pore presence. These can maintain a certain contamination, which is difficult to remove and will not allow the release of decontaminated materials into the environment.

Solid RAW (from decontamination and secondary RAW) was processed by incineration in the nuclear facility "RAW Bohunice treatment and conditioning centre". For this the RAW was packed in Tatrafol bags, with standard dimensions of 550 x 650 x 0.08 mm and then transported in 200 L MEVA drums. About 500 kg of dried DeconGel, whose volumetric activity β γ did not exceed 6×10^6 Bq/kg, was created from 4000 m^2 surface area of the tanks.

The Fig. I–3 presents the facility in at the premises of V1 NPP after decommissioning of the collection tank and decontamination using DeconGel.

FIG. I–3. Decontaminated facility available for processing Ra-contaminated solid material. (Courtesy of JAVIS, Slovakia).

ANNEX II
INTERNATIONAL STUDIES ON WET SOLID WASTE MINIMIZATION

The examples provided in the following pages address the organization and detailed technical aspects of examples of international case studies on wet solid waste minimization at NPPs. The descriptions provide practical information on projects are planned and managed in various Member States. The examples given in this Section are not necessarily best practices but rather reflect a variety of international practices. These case studies reflect the experience and views of their contributors and, although consistent with the main text, are not for guidance.

II–1. FRANCE CASE STUDY

Cyclife France is an EDF Devoted Waste treatment Subsidiary from EDF Group. Located in France, it provides radioactive waste treatment services directly on production sites thanks to mobile treatment and packaging units operated by its expert staff. To meet the most complex requirements, Cyclife France has designed, manufactured and commissioned mobile units which Cyclife temporarily set up and operate directly at customers' radioactive waste producing sites (Fig. II–1).

FIG. II–1. MERCURE mobile facility arriving on site. (Courtesy of EDF, France).

Each mobile machine packages diverse types of waste. For the conditioning of ion exchange resins (IER), Cyclife France developed a specific process: MERCURE.

The wastes that can be processed by the MERCURE process are:

— Chemical streams containing high concentrations of borates, lithium, iron, cobalt, nickel, chromium, sodium and calcium will be accepted;

— Ion exchange resins with a base structure of polystyrene, phenolic, acrylic or formophenolic that are in the form of beads or grains with a diameter of between 0.3 and 1.2 mm;

— Cationic or anionic resins;

— Radionuclides accepted: ^{58}Co, ^{60}Co, ^{110m}Ag, ^{54}Mn, ^{57}Co, ^{134}Cs and ^{137}Cs;

— Radiological criteria: total specific activity of βγ emitters < 13,500 GBq/m^3.

The process used by Cyclife France on the MERCURE machines coats the IER with an epoxy resin matrix (Fig. II–2).

FIG. II–2. MERCURE Process in operation at production site. (Courtesy of EDF, France).

The waste form is packaged in a concrete container, reinforced with integrated steel radiation shields (Fig. II–3), manufactured at Centraco.

FIG. II–3. MERCURE shielded container. (Courtesy of EDF, France).

Since their commissioning in 1996 and 2002, the MERCURE machines have packaged almost 5 000 m^3 of IER from PWR primary circuit.

MERCURE processing provides radioactive waste treatment possibilities directly on production sites as Cyclife France have designed, manufactured and commissioned mobile units which Cyclife temporarily set up and operate directly at customers' radioactive waste producing sites.

II–2. SOUTH AFRICAN CASE STUDY

During the purification of uranium for the manufacturing of Mo target plates, different waste streams are generated that could contain substantial amounts of uranium that needs to be recovered in order to dispose these effluents (Fig. II–4).

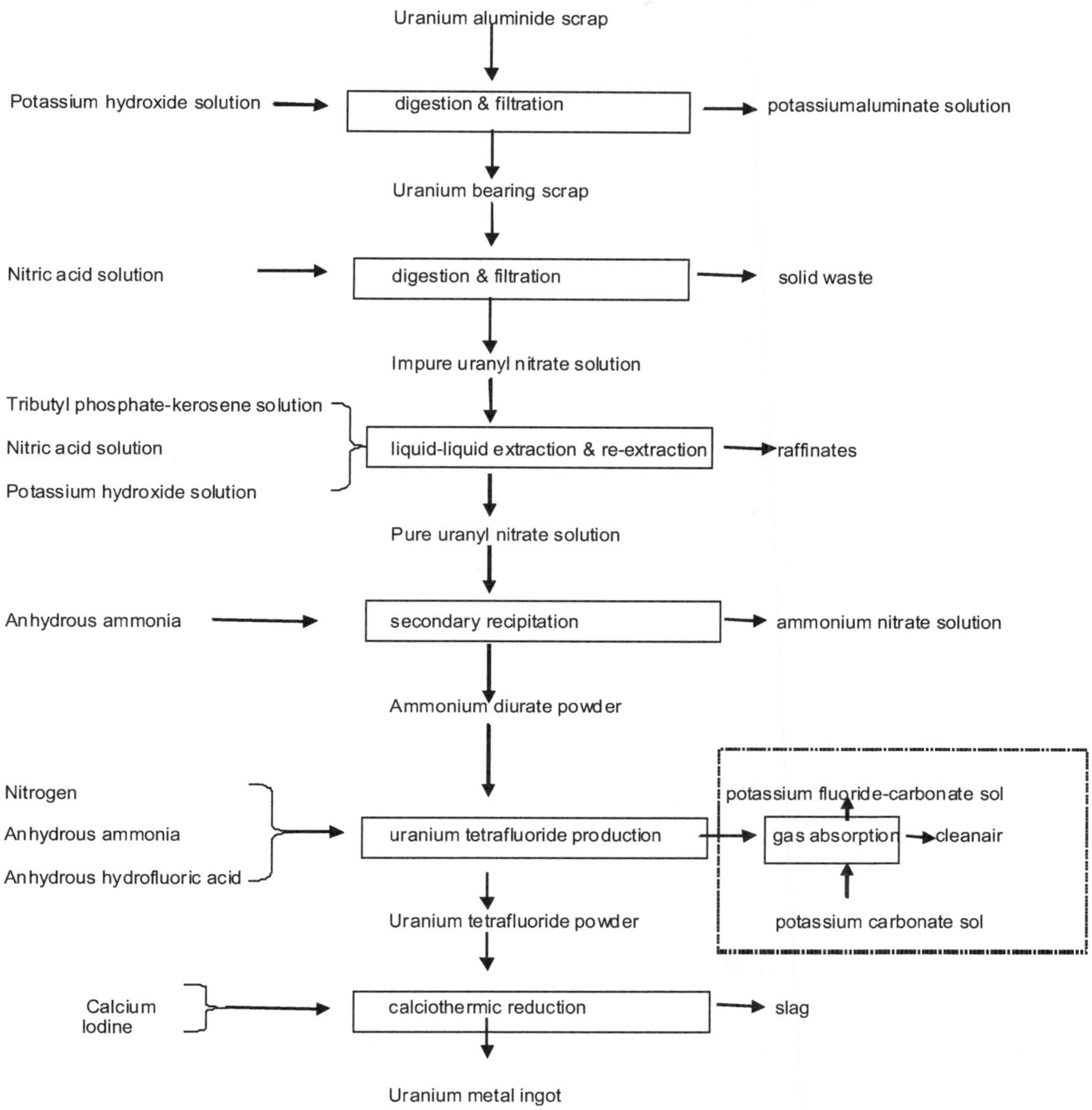

FIG. II–4: Schematic presentation of uranium purification methods employed by South African Nuclear Energy Corporation (NECSA) to purify uranium for target plate manufacturing.

The different waste streams can be summarised as indicated in Table II–1.

Analysis of these waste streams indicated that the concentration of uranium present exceeds the disposal requirement of 1 Bq/l. The uranium could be removed from the different waste streams using various technologies, but best case would be if uranium could be recovered using minimal different ion exchange media.

TABLE II-1. DECRIPTION OF GENERATED WASTE SOLUTIONS

Waste stream	Origen/ Source	Main Composition	Liquid waste tank
W06 Filtrate	Filtrate from the dissolution of aluminium scrap	KOH; $KAl(OH)_4$	WT02, WT05
W06 Raffinate	Raffinate from the batchwise tributyl phosphate-kerosene liquid extraction	20% HNO_3	WT03
RPU	Raffinate from the continuous tributyl phosphate-kerosene liquid extraction plant	20% HNO_3; $Al(NO_3)_2$	WT04
FKP	Filtrate from the precipitation of Ammonium diurinate	3.5 % mass NH_4NO_3	WT09
SWL/MWL	Filtrate from the leaching of iodine from Calcium fluoride; iodine slag and magnesite	CaI_2;CaF_2;I_2	WT06, WT07
FWS	Floor wash solutions	Soap	WT01
SOP	Spent solutions from hydrogen fluoride furnace scrubbers	NH_3;K_2CO_3;KF	WT08

No user requirements were specified regarding the investigation of the removal of uranium from the different waste streams, but the following was considered:

— The technology needs to be able to be accommodated into the current process;

— The technology needs to be able to be accommodated by the current licence conditions;

— No gaseous releases are allowed;

— Minimum additional chemicals to be implemented, to satisfy current licence requirements;

— Criticality needs to be considered, thus ensuring that the accumulation of uranium is limited, managed, or contained in geometric criticality-safe containers or tanks.

The choice of resin depends on the capacity of the resin, The K_d value of the resin and the waste stream composition are indicated in Table II–2.

TABLE II-2. EXAMPLES OF ION EXCANGE RESINS USED IN NUCLEAR INDUSTRY

Ion exchange resin	Operating conditions
Diphonix	Cation exchange Acidic medium. 0.1M HNO_3 Convert to basic medium with 0.5M KOH Leach uranium with 2M Na_2CO_3 / 5% H_2O_2 at pH=9
Bio-Rad AG 1x8	Anion exchange 8M HCℓ Leach uranium with 1M HCl
AG1-X4 (Converted to MnO_2 impregnated resin)	Anion-exchange (Cl-form), neutral pH
Birnessite	Cation-exchange Manganese oxide material pH = 4
Amberjet 4400 (Cl-form)	Basic (carbonate) media Anion gel-type pH: 9.2 – 10.4
Amberlite IRA 910U (Cl-form)	Anion exchange resin Carbonate medium Leaching with NH_4NO_3 or NaCl pH: 8 – 10
Bio-Rex 5	Anion exchange Strong basic medium 1M Na_2CO_3 medium CO_3^{2-} / HCO_3^- 0.18M/0.02M pH : 7–11
Dowex 1-X4 (Cl-form)	1M H_2SO_4 (convert to SO_4^{2-}) Wash with water

Initially batch experiments were performed using different ion exchange resins to remove uranium from the different waste tanks and the 'best' results per batch experiment were continued using column experiments to determine the capacity, recovery and break though curves. The results from this study (Table II–3) indicated that only two ion exchange resins will remove sufficient uranium (for release) from the different solutions generated during the manufacturing of Mo target plates.

TABLE II-3. ION EXCHANGES THAT COULD RECOVER U FROM DIFFERENT LIQUID WASTE STREAM COMPOSITIONS (FOR FREE RELEASE).

Waste Stream	Origen/ Source	Main Composition	Liquid waste tank	Ion exchange
W06 Filtrate	Filtrate from the dissolution of aluminium scrap	KOH, $KAl(OH)_4$	WT02, WT05	Amberlite IRC747 Purolite S940
W06 Raffinate	Raffinate from the batchwise tributyl phosphate-kerosene liquid extraction	20% HNO_3	WT03	Amberlite IRC747 Purolite S940 Diphonix
RPU	Raffinate from the continuous tributyl phosphate-kerosene liquid extraction plant	20% HNO_3, $Al(NO_3)_2$	WT04	Amberlite IRC747 Purolite S940 Diphonix
FKP	Filtrate from the precipitation of Ammonium diurinate	3.5 % mass NH_4NO_3	WT09	Amberlite IRC747 Purolite S940 Diphonix
SWL/MWL	Filtrate from the leaching of iodine from Calcium fluoride; iodine slag and magnesite	CaI_2, CaF_2 ,I_2	WT06, WT07	Not evaluated
FWS	Floor wash solutions	Soap	WT01	Diphonix
SOP	Spent solutions from hydrogen fluoride furnace scrubbers	NH_3, K_2CO_3, KF	WT08	Amberlite IRC747 Purolite S940

II–3. UNITED STATES CASE STUDIES

II–3.1. Diablo Canyon demineralizer vessel short loading

Diablo Canyon is a two-unit Westinghouse PWR. Each unit has one 850 L demineralizer vessel to treat the spent fuel pool. Each unit also has a similarly sized vessel to treat reactor coolant and another for shutdown cleanup. The original load for each of these three vessels was a stoichiometric mixed bed: two parts anions to one part cation. These beds always exhausted on the cation component, never on the anion component. To reduce resin procurement and disposal costs, the anion load was gradually reduced. The rule of thumb to prevent channelling through a resin component is that 0.3 m of bed height is needed. For the online letdown lithiated mixed bed, the anion component was reduced to 0.3 m bed height, which equalled 142 L for the 0.76 m diameter vessel, with no adverse effect on reactor coolant water quality. The original load of 283 L of lithiated cation resin remained unchanged. The total bed volume was reduced by 50%, from 850 L to 425 L, simply by loading less anion resin. These beds last for an extended 21-month fuel cycle with zinc addition, with no adverse effect on water quality. In fact, water quality was improved because chloride levels are reduced due to the lower anion resin content in the bed.

A similar change in the load was made for the shutdown clean up bed. This load was reduced from 850 L to 452 L by loading less anion resin. However, a 142 L layer of macroporous anion resin was loaded on top of this bed to remove particulate and colloid activity liberated during the reactor shutdown. This not only reduced the number of downstream cartridge filters generated during the shutdown but, also allowed the bed to be used for two reactor shutdowns versus one.

For spent fuel pool service, the anion component of the mixed bed was likewise reduced from 566 L down to 142 L. Water in the presence of neutron flux in the pools could hydrolyze forming hydrogen peroxide (H_2O_2) which will cleave the bonds in cation resin resulting in resin breakdown after about 12 months of service. The cation load was reduced over time to try to match the resin volume with resin life. The cation resin was still breaking down before exhausting on cobalt with the minimum 142 L cation load. This resulted in a 283 L bed. However, due to elevated levels of $^{124/125}Sb$ in the pools from Vantage V Westinghouse fuel and ^{63}Ni from old fuel with Inconel grid straps, a 142 L layer of selective resin is loaded in the vessels to remove these nuclides. This results in a 425 L bed which is half the original load with no adverse effect on water quality or bed life.

II–3.2. Diablo Canyon polymer solidification of bead resins

In the USA, intermediate level waste (Class B and C) resin needs to be stabilized to Nuclear Regulatory Commission (NRC) criteria. Such spent resin can either be dewatered in a high integrity container (HIC) or solidified with a binder that can pass various waste form tests to provide stability. Although LLW disposal sites in the USA accept HICs, the proposed California LLW disposal site did not. Diablo Canyon evaluated solidification services to comply with the proposed disposal site. Cement solidification of spent resin would increase the waste volume and sometimes had curing issues. A new polymer solidification process had been developed to stabilize spent resin with chelating agents from chemical decontamination processes. The US NRC had approved this process for stabilization and a batch of spent resin from an EPRI fuel decontamination project had been packaged using vinyl ester styrene (VES). This process offered the hope of stabilizing resin with no increase in waste volume.

A container of Diablo Canyon resin was solidified with VES to determine if the process would be suitable for the proposed California disposal site. It was determined in specimen testing that the spent resin was not fully exhausted. The VES binder could not fully flow through the resin unless it was first exhausted. A liquid chemical additive had to be metered into the container to exhaust the resin. The resin container was successfully solidified but, a secondary chemical liquid waste was created which had to be treated off-site. The vendor understood this secondary waste stream would not be acceptable to clients going forward and began working on a different binder formulation.

Although the California LLW disposal site was granted a license, a new State Governor opposed the project, and the site was never opened. Diablo Canyon's access to disposal of Class B and C waste at a LLW disposal site in South Carolina was to end in 2008. The plant had been forced to store waste on-site when access to disposal was lost for one year in the 1990's. Dewatered spent resin had been packaged in a poly HIC for stability inside a carbon steel container for fire protection. Due to the foggy coastal environment for the plant, after storage, waste containers had to be re-dewatered prior to shipment off-site. This method of packaging for storage was costly because two waste containers had to obtained for each storage location (poly HIC inside steel container). It was also dose intensive because the resin had to be dewatered twice on-site.

The solidification vendor developed a new polymer for the U.S. Navy that did not require spent resin to be chemically treated before binder addition. Diablo Canyon used the outer carbon steel storage container to package resin with the Advanced Polymer process. This process did not increase waste volume. In fact, the amount of resin that could be stored increased because the outer steel container had a greater capacity than the poly HIC.

By storing the polymer solidified waste in the steel containers with their lids off, moisture did not condense in the containers and no dewatering post storage was required. This reduced radiation exposure from spent resin requiring storage on-site.

In 2012, a new Disposal site opened in Texas. By 2018, Diablo Canyon had shipped all the waste that had to be placed in storage since 2008. On-site storage of polymer solidified spent resin for a decade was achieved without increasing waste volume while limiting any increase in radiation exposure.

II–3.3. Diablo Canyon use of filter shear

Packaging of spent cartridge filters is very inefficient. Placing one filter in a 200 L cement lined drum yields a packaging efficiency of only 13%. Diablo Canyon uses a 2.27 m^3 carbon steel container with an internal cage for the encapsulation of spent filters. Encapsulation ensures the 'no free water' disposal criteria is met.

The capacity of the internal cage is about 1.7 m^3. Diablo Canyon has three types of filters. The random loading of spent filters into the large waste container didn't improve the packaging efficiency very much above the one filter per drum case.

Diablo Canyon implemented a filter shear system in conjunction with a spent filter transfer cask to optimize radioactive waste packaging efficiency. The shear is positioned over the port of the packaging shield, allowing the transfer cask to be securely placed on top. During the process, most spent filters undergo one or two shear operations as they are lowered from the transfer cask into the waste container. This method significantly enhances packaging density, enabling approximately four times the number of sheared filters to be loaded into a waste container compared to unsheared filters. By increasing container utilization and reducing the volume of waste shipments, this approach contributes to improved waste management efficiency, cost savings, and compliance with regulatory disposal requirements.

ANNEX III
INTERNATIONAL STUDIES ON SOLID WASTE MINIMIZATION

The examples provided in the following pages address the organization and detailed technical aspects of examples of international case studies on wet solid waste minimization at NPPs. The descriptions provide practical information on projects are planned and managed in various Member States. The examples given in this Section are not necessarily best practices but rather reflect a variety of international practices. These case studies reflect the experience and views of their contributors and, although consistent with the main text, are not for guidance.

III–1. FRANCE CASE STUDIES

III–1.1 Waste minimization training at EDF industrial demonstrator

The complexity of the decommissioning process for graphite reactors is due to their dimensions, geometry and access as well as the presence of irradiated graphite itself. Dismantling activities and waste generated from graphite reactors are ten to twenty times greater than those of other nuclear technologies. Limited operational experience is available worldwide on this topic. Indeed, up to now, only small research graphite reactors (Fort Saint Vrain (USA), Brookhaven (USA) and Windscale AGR (UK) have been dismantled.

Such complexity raises four types of risks. Three are related to the graphite reactors design, such as concrete and metallic structures cutting operations and graphite brick retrieval operations. The fourth one is related to remotely operated tools.

It is therefore important to implement a progressive risk mitigation approach to achieve safe and optimized dismantling activities of a first of a kind graphite reactor and reduce waste production. This approach includes on site investigations in addition to technological and digital development taking place in a dedicated facility.

In this context, EDF decided to design and build an Industrial Demonstrator (Fig. III–1) in which technological developments including processes and tests using remotely operated tools developed by Graphitech (an EDF and Veolia joint venture) in the testing hall (Fig. III–2) is being evaluated. The expected outcomes of these tests and developments will enable EDF to start the dismantling activities of Chinon A2 graphite reactor by 2030.

FIG. III–1. Industrial demonstrator. (Courtesy of EDF, France).

FIG. III–2. Industrial demonstrator testing hall. (Courtesy of EDF, France).

To achieve an optimized and safe decommissioning of Chinon A2, the following is being evaluated in the Industrial Demonstrator:

- — Development of remotely operated tools;
- — Development of digital solutions;
- — Development of training modules.

III–1.2. Development of remotely operated tools

Development of modern technologies and techniques conducted by EDF group through its subsidiary Graphitech to provide the best solutions in terms of safety and radiological protection for remote dismantling operations. A full-scale mock-up which are representative of the Chinon A2 reactor can be seen in Fig. III–3.

FIG. III–3. Full scale mock-ups evaluated in Industrial Demonstrator (ID). (Courtesy of EDF, France).

III–1.3. Development of digital solutions

Different advanced digital solutions to support the remotely operated and robotic developments and test programme have been developed. For example, DEMplus for nuclear software, developed by Cyclife Digital Solutions (another EDF group subsidiary) will simulate the dismantling scenario in 3D to optimize modelling and analyse the impact of different alternatives of a scenario. A 3D modelling of Chinon A2 NPP is presented in Fig. III–4. These simulations give an estimation of the schedule, the dose rate received by the workers, the waste production and the dismantling cost. The outputs of such simulations will help optimizing dismantling scenario as well as waste management and minimization.

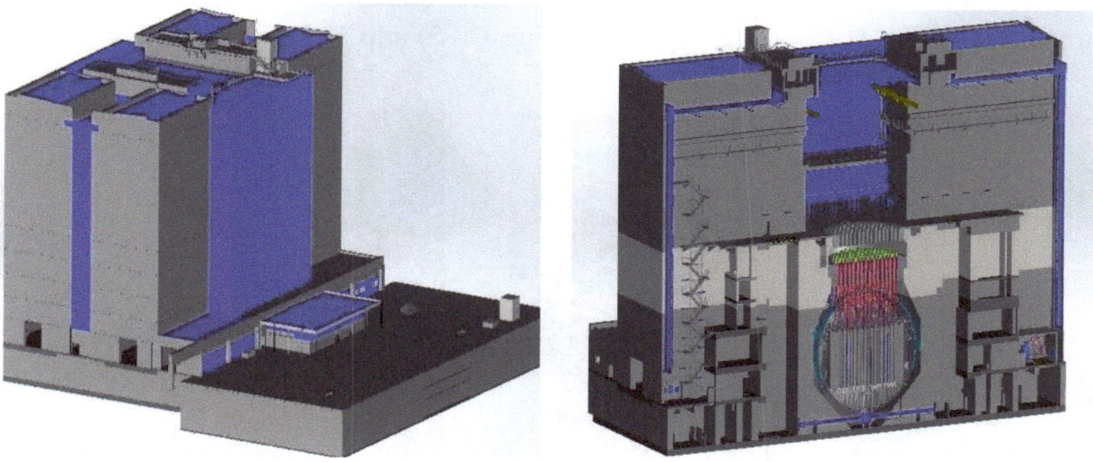

FIG. III–4. 3D modelling of Chinon A2 NPP. (Courtesy of EDF, France).

III–1.4. Development of training modules

Training modules will be available for operators, project managers and engineers working on decommissioning scenarios as well as for anyone wishing to learn more about graphite reactors decommissioning and remotely operated solutions. For example, in addition to in-class training sessions, the operators will receive training on remote dismantling operations using digital solutions.

Finally, applying lessons learned from these operations will enable EDF and operators worldwide to adjust and improve the equipment and tools, optimizing the dismantling process and reducing the waste produced during such operations. This will benefit all French graphite reactors and their peers worldwide.

The knowledge gained will facilitate international cooperation and information exchange in decommissioning projects, including harmonization of practices and associated standards will ensure waste minimization through training.

III–1.5. Waste minimization of contaminated metal by melting at Cyclife

One process to minimize waste volumes is the decontamination of metal surfaces (e.g., steam generators) by melting for declassification the final metal ingot as very low-level waste for storage or disposal. Melting of metals can have different purposes, including:

— Minimizing the disposal volume of metals;

— Homogenization of the material for disposal;

— Simplifying and obtaining more precise measurements of activity in the material;

— Clearance and recycling of metal outside the nuclear industry.

Cyclife operates three facilities dedicated to the treatment of low-level radioactive contaminated metals generated by nuclear industry, including two facilities operating melting furnaces (CENTRACO facility in France (Fig. III–5) and Nykoping facility in Sweden (Fig. III–6).

FIG. III–5. Schematic layout of the melting facility (Cyclife France plant) in France. (Courtesy of Cyclife-EDF, France).

FIG. III–6. The metal melting facility of Cyclife Sweden (Courtesy of Cyclife-EDF, France).

The combined capacities of the facilities are between 8500 tonnes of contaminated metals yearly. Accepted metals are ferrous and non-ferrous material: aluminium, copper lead and special alloys. Contaminated metals with up to 100 Bq/g initial activity can be subject to clearance after treatment (Nyköping facility).

The pretreatment line of melting unit consists of the following main components (located in controlled area):

- A reception and sorting workshop where waste to melt is unloaded;
- A manual plasma torch workshop;
- An industrial automatic shear for size reduction of thin metallic parts;
- A separate workshop dedicated to large components where those parts are dismantled and size reduced by remoted operations (sodium cold traps, used fuels racks, heat exchangers). It is composed of:
  - An automated diamond wire cutting saw;
  - An automated heavy-duty shear;
  - An automated heavy-duty shredder;
  - An automatic ferrous and non-ferrous metal sorting line;
  - A fleet of reusable boxes to store the batches of prepared metals.

The line of melting unit consists of the following main components are:

- An automated cinematic line to bring the filled boxes to the furnace level where they are emptied in a hopper connected to a vibrating conveyor to feed the furnace;
- A 4 tons capacity melting induction furnace;
- Static moulds to cast 1.5 t ingots;
- A centrifuge machine to recycle part of the waste by moulding cylindrical shields;
- A HVAC and off-gas filtration consisting of a bag-house filters;
- A radiological and chemical off-gas monitoring unit connected to the stack of the incinerator;
- A control room in non-controlled area to drive and monitor all operations for melting.

The shipping of EDF Fessenheim PWR NPP Steam generators from France to EDF-Cyclife Facility in Sweden for melting can be seen in Fig. III–7.

FIG. III–7. Shipping of EDF Fessenheim PWR NPP Steam generators from France to EDF-Cyclife Facility in Sweden (Courtesy of Cyclife-EDF, France).

On arrival at the Studsvik facility the boilers were subject to radiological surveys, to identify radiological conditions and collection of any non-fixed materials. Segmentation followed before melting.

The project has demonstrated that large boilers can be safely transported and effectively decontaminated to enabling valuable metal to be recycled (up to 96%) within a short timescale. Secondary waste that arises during melting is suitable for disposal as LLW.

III–2. INDIA CASE STUDY

III–2.1 Alternative metals to reduce neutron activated impurities.

The Indian Pressurized Heavy Water Reactors (PHWRs) programme consists of 220 MWe, 540 MWe and 700 MWe units. At present India is operating 16 units of 220 MWe. The detail system design of 220 MWe includes:

- Reactor regulating system consists of:
  - Automatic liquid poison addition system (ALPAS) is included to facilitate the reactor output;
  - For regulating of power: 4 cobalt/stainless steel regulating rods;
  - For quick power reduction: 2 cadmium sandwiched stainless steel shim rods;
  - For xenon override during the regulations of power: 8 cobalt/stainless steel absorber rods to provide xenon override capability.
- Reactivity in PHWRs is controlled by:
  - Regulating rods or central adjusters for the manipulation of reactivity;
  - Absorber rods or corner adjusters to compensate the burn-up load during non-availability of fuelling machine;

  - Shim rods as backups to regulating rods to ensure negative reactivity during power declines.
- Shutdown systems:
  - Primary shutdown system consists of 14 mechanical rods of cadmium sandwiched between stainless steel elements;
  - The secondary shutdown system provides fast filing of liquid poison in 12 vertical tubes located inside calandria.
- Cobalt absorbers rods to produce ^{60}Co for commercial applications is part or reactor system.

To minimize waste arisings the following strategies are used:

- Reducing equipment drains by using better seals, leak free joints and proper monitoring methods;
- Optimum use of Ion exchange columns in the purification system;
- Controlling issue of material used in active areas;
- Painting the wooden sleepers for easy decontamination; Removing packing materials outside the active area;
- Improving housekeeping;
- Keeping all the protective gears at their designated bins;
- Ensuring proper planning of maintenance work;
- Ensuring careful movement of radioactive material;
- Optimum use of hand gloves.

III–2.2. Pressure tubes replacement

Pressure tubes in pressurized heavy-water reactor (PHWR) encounters high neutron fluxes, temperatures and high-pressure during operation. Initially pressure tubes constructed from Zircaloy-2 material were used at RAPS-2, MAPS-1, MAPS-2, NAPS-1, NAPS-2 and KAPS-1. The composition of the pressure tubes was modified to Zr-2.5% Nb to reduce the wall thickness increasing neutron economy gains. The activated radionuclides were ^{94}Nb and ^{60}Co and the presence of ^{94}Nb (half-life of 20 300 years, two gamma rays of 703 and 871 KeV energy requiring long term isolation. Pressure tubes (Zr-2.5% Nb) has been used in reactors (KAPS-2 onwards) and is already included in all modern designs.

III–2.3. Reuse of filters

The ventilation exhaust system of the reactor building, reactor auxiliary building (RAB) and waste management plant (WMP) use pre-filters that needs to be replaced when reaching the pre-defined differential pressure across the filters. To minimize waste, the pre-filters are removed and washed by a high-pressure jet in the reverse air flow direction inside a controlled area. Pre-filters can be reused for up to three cycles.

Spent HEPA filters are removed, metallic filter frame cut open and the filter media placed into a 200–litre drum before compacting for further volume reduction. The metallic filter frames are decontaminated and certified as inactive metallic scrap.

III–2.4. Reducing ^{60}Co activity

The problem of hot particle spots by deposition of radioactive particles was experienced in Indian PHWR stations. Hot spots are observed on the dead ends of the piping of moderator and the primary heat transport system. In the NPPs, cobalt ingress in moderator system from cobalt-based component of regulating rods has been experienced. During initial days of operation, the release of cobalt from these components was not realised and purification columns were valved out during shutdown. Logics and procedures were modified so that boron saturated IER columns are only valved in during unit shutdown to ensure removal of cobalt activity. This has reduced the Cobalt activity build-up and hot spot.

With ^{60}Co arising (long-lived activation product) selecting materials containing low cobalt concentrations for the primary heat transport system and moderator systems is important to reduce arises. Also, to the maximum extent possible, for the minimising of cobalt bearing materials (e.g., stellite for valve seats, bearing, materials, components and piping), these materials are currently being replaced in NNPs with cobalt free substitutes.

III–3. SLOVENIA CASE STUDY

Slovenian legislation prescribes waste management programme for all radiation or nuclear facilities. The radioactive waste management programme, in addition to other prescribed contents, also contain technical, organizational and other measures taken to reduce the generation of RAW. According to legal requirements, Krško NPP prepared the general document “Radioactive waste management programme” which is reviewed at least every two years while technical solutions are described in technical report ‘Radioactive waste management in Krško NPP’. The technical report is an internal document in which the existing practice of handling RAW in the Krško NPP are described and based on experience and development in this field. The trend and forecast of the filling of low and medium RAW storage facility in the Krško NPP in the past and the projection for waste from new projects are given. The technical report serves as a basis for long-term planning of continuous and campaign processes, such as super compaction, incineration and melting, and identification of the need to introduce additional processes for safe and rational storage and optimization of costs.

Additionally, the internal document: ‘Internal policies and objectives - five-year development plan’ provides direction on the reduction of the amount of RAW by consistent control and elimination of coolant leaks, decontamination and reuse of tools and materials, prevention of contamination, consistent separation of contaminated materials from clean ones and the use of volume reduction techniques. Krško NPP initiated the following actions to decrease the generation rate of radioactive waste of various types:

- — Melting of metal radioactive waste;
- — Carrying out super compaction campaigns to reduce the volume of the solid radioactive waste to be stored;
- — Incineration of combustible radioactive waste;
- — Metal materials exceeding exemption/clearance levels are stored onsite before melting;

— Consistent control of the entry and exit of materials into the radiologically controlled area;

— The use of materials that are easier to decontaminate (stainless steel instead of wood/plastic);

— Segregation techniques are used for collecting non-contaminated materials separately and to allow waste streams to be processed separately.

Segregation of RAW are based on characteristics, (e.g., compressible, combustible, un-compressible, wet rags, oily rags) as illustrated in Fig. III–8. RAW type is distinguished by use of collection bags in assorted colours. Coloured bags are also used to separate equipment and tools being used in a controlled area. Green bags are used for noncontaminated and yellow bags for contaminated equipment that can be decontaminated. Red bags are used for equipment to be put in waste.

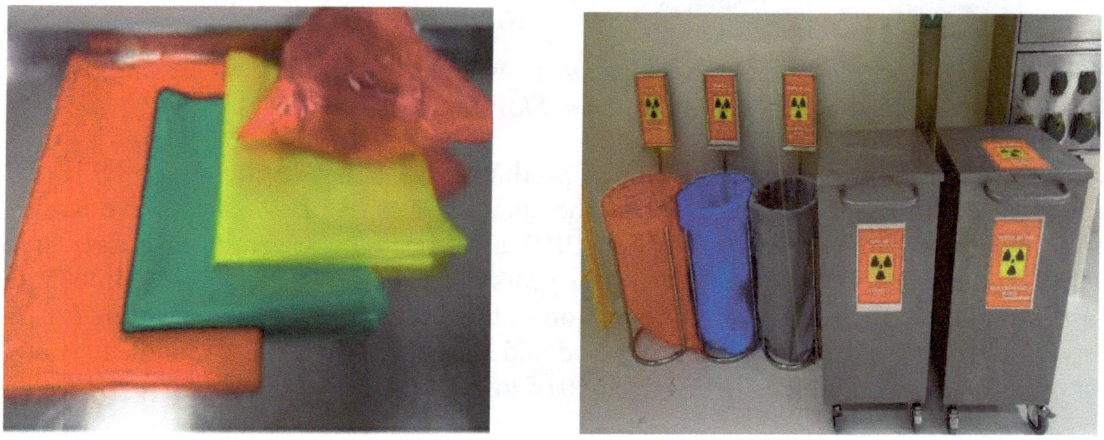

FIG. III–8. Segregation of RAW at Krško NPP. (Courtesy of Nuclear Safety Administration, Slovenia).

To minimize resin volumes and evaporator bottoms (concentrate) the original Westinghouse treatment procedure (encapsulation) was replaced with an In-drum Drying System. For spent resins, this system the generated free-flowing resin beads could be filled directly into 200 L stainless steel heavy drums with biological shields (150 L of usable volume). Treated secondary spent resins (with minimal radioactivity) are filled into 200 L stainless steel drums without biological shielding. Evaporator bottoms and sludges during drying converts into dry solid waste products with no free water and low residual moisture (Fig. III–9).

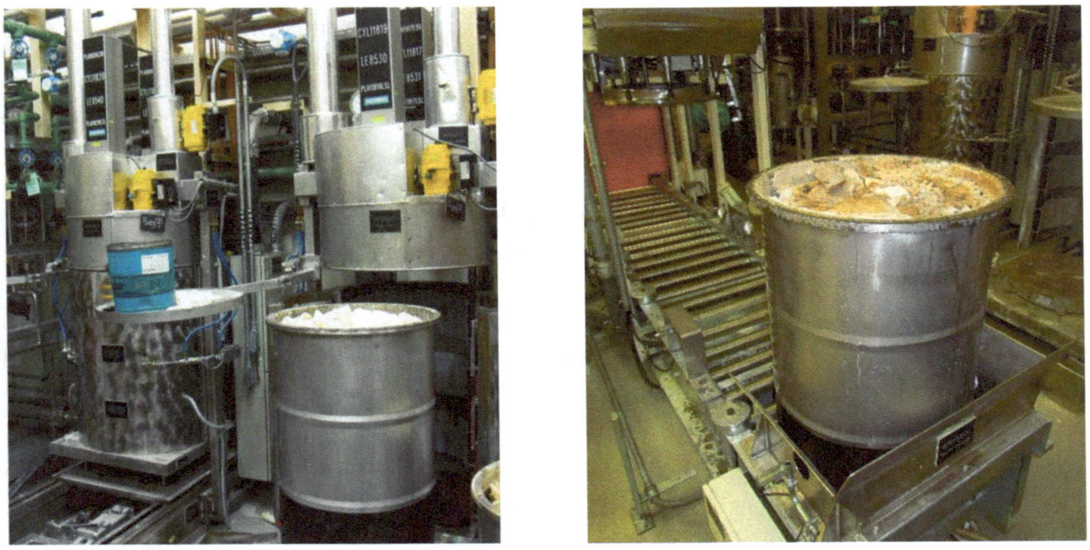

FIG. III–9. In-drum Drying System to reduce resin volumes before processing at Krško NPP. (Courtesy of Nuclear Safety Administration, Slovenia).

With the above improvements the annual production of LILW at the Krsko NPP declined as indicated in Fig. III–10. Figure presents the annual arises of LILW (dry active waste, sludge, resins and evaporator bottoms) at Krško NPP that has been processed for disposal. The increase in the last three years compared to previous years is due to the inaccessibility of the radioactive waste storage facility during the construction of the waste manipulation building (2016–2018). The RAW from these years were processed and stored in 2019, 2020 and 2021. However, the last three years also includes RAW from 2016 to 2018.

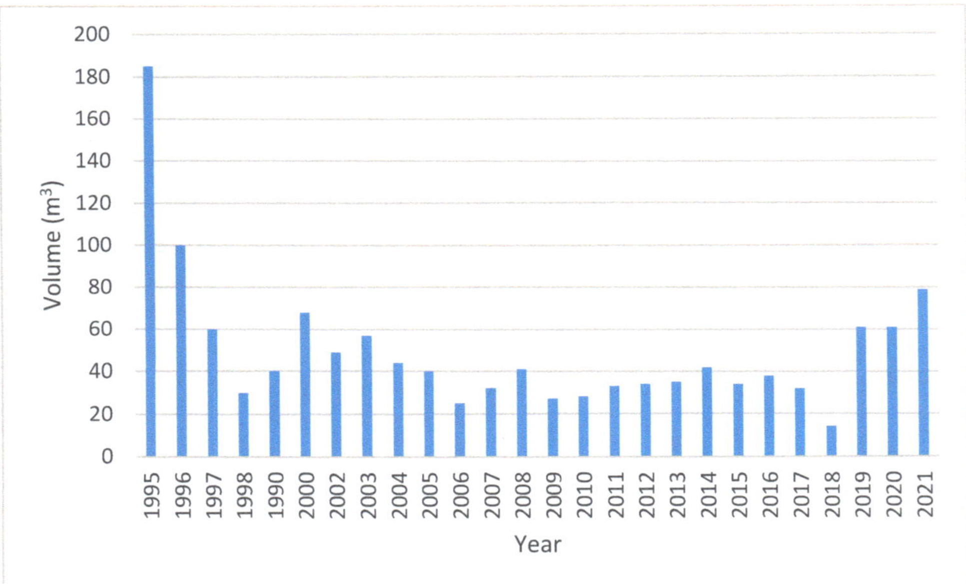

FIG. III–10. Annual arises of LILW at Krško NPP. (Courtesy of Nuclear Safety Administration, Slovenia).

III–4. UNITED STATES CASE STUDIES

III–4.1. Diablo Canyon use of launderable rubber garments

Diablo Canyon, a 2-unit PWR, was provide with an on-site laundry for radioactively contaminated garments. Dry cleaning units were provided for cloth garment and wet washers and clothes dryers for rubber garments. By the time the steam generator replacement projects were implemented for both units, the dry cleaners had been decommissioned and dissolvable garments were in use versus cloth garments. Re-usable rubber garments remained in use. The Steam Generator Project Contractor demanded disposable rubber garments be used. They wanted the dress out area in the Containment Access Facility (CAF) to be radiologically clean.

The mass of the 'single use' rubber gloves and shoe covers procured by the project was compared that figure to the total mass of contaminated rubbish arising from each unit project. In both cases the disposable rubber garments made up 50% of the contaminated waste.

Based on this experience, launderable rubber garments were used for the subsequent Reactor Pressure Vessel Head replacement projects. The cost to operate the on-site laundry with the inventory of launderable rubber garments already on-site was far less than procuring 'single use' rubber garments and paying to dispose of them as radioactive waste.

III–4.2. Canyon use of launderable fire-retardant garments

During the Steam Generator Replacement projects at Diablo Canyon, increased use of fire–retardant garments in contaminated areas was required to support welding activities and new arc flash protection criteria. Disposable fire-retardant plastic coveralls were available but, there were no shoe covers or hoods. Special fire-retardant treated cloth shoe covers and hoods were procured. It was found that the plastic coveralls did not breath and became hot and were very uncomfortable for the workers. The fire-retardant cloth garments were comfortable and were launderable.

For the reactor vessel head replacement projects, fire retardant cloth coveralls were obtained. These proved comfortable. Although dissolvable fire-retardant garments are now available, the cloth garments are more comfortable and generate less radioactive waste as they can be laundered and reused.

ABBREVIATIONS

Bq	becquerel
Bq/g	becquerel per gram
CA	contaminated areas
CFAM	corporate functional area manager
LLW	low level waste
MLR	mid-life replacement
NNSA	National Nuclear Safety Administration
NPP	nuclear power plant
PE	polyethylene
PPE	personal protective equipment
PT	pressure tubes
PU	polyurethane
PVA	poly-vinyl-alcohol
PVC	poly-vinyl-chloride
PWR	pressurized water reactor
RAFM	reduced activation ferric or martensitic
RAW	radioactive waste
RCA	radiological controlled area
RO	reverse osmosis
WANO	World Association of Nuclear Operators
WWER	water-water energetic reactor

CONTRIBUTORS TO DRAFTING AND REVIEW

Amyot, C.	Ontario Power Generation, Canada
Andrei, V.	Societatea Nationala Nuclearelectrica, Romania
Banach, P.	Polskie Elektrownie Jądrowe, Poland
Bhachawat, D.	Nuclear Power Corporation, India
Bugala, E.	Polska Grupa Energetyczna, Poland
Campbell, D.	Electric Power Research Institute, United States of America
Dragolici, F.	International Atomic Energy Agency
Farkas, P.	Nuclear Safety Administration, Slovenia
Forsberg, R.	Loviisa Nuclear Power Plant, Finland
Hrkut, M.	Slovenské elektrárne, Slovakia
Huichun, L.	Sanmen Nuclear Power Co. Ltd, China
Janski, S.	International Atomic Energy Agency
Khan, A.W.	Chashma Nuclear Power, Pakistan
Kukan, T.	Nuclear and Decommissioning Company, Slovakia
Lange, K.	International Atomic Energy Agency
Malatim, T.	Atomic Energy Establishment, Libya
Meyer, W.C.M.H	International Atomic Energy Agency
Miller, C.	Pacific Gas & Electric Company, United States of America
Narayanan, R.A.	Nuclear Power Corporation, India
Ötvös, N.	Atomic Energy Authority, Hungary
Pieraccini, M.	Électricité de France, France
Rakitskaya, T.	ROSATOM State Corporation, Russian Federation
Sampaio J.V. J	Eletrobras Eletronuclear, Brazil
Shivaji Krishna Pawar, S.	Atomic Energy Regulatory Board, India
Stancheva, V.	Kozloduy NPP, Bulgaria
Stein, M.	Kernkraftwerk, Leibstadt, Germany
Virtopeanu, C.S.	National Commission for Nuclear Activities Control, Romania

Consultants Meetings

Vienna, Austria: 24–31 May 2021, 18–22 July 2022, 04–08 Dec 2023

Technical Meeting: Vienna, Austria: 22–26 November 2022

IAEA
International Atomic Energy Agency

CONTACT IAEA PUBLISHING

Feedback on IAEA publications may be given via the on-line form available at:
www.iaea.org/publications/feedback

This form may also be used to report safety issues or environmental queries concerning IAEA publications.

Alternatively, contact IAEA Publishing:

Publishing Section
International Atomic Energy Agency
Vienna International Centre, PO Box 100, 1400 Vienna, Austria
Telephone: +43 1 2600 22529 or 22530
Email: sales.publications@iaea.org
www.iaea.org/publications

Priced and unpriced IAEA publications may be ordered directly from the IAEA.

ORDERING LOCALLY

Priced IAEA publications may be purchased from regional distributors and from major local booksellers.

25-01268E